THE PONY TRAP

TO GAY, ABBEY, DUSTY AND PAPI
Without whose endearing habits, this could have been
written in half the time.

THE PONY TRAP

Pamela Donald

WEIDENFELD AND NICOLSON ° LONDON

Title page: illustration by Deborah Hall (age 14)

Copyright © 1987 Pamela Donald

Published in Great Britain by
George Weidenfeld & Nicolson Limited
91 Clapham High Street
London SW4 7TA

ISBN 0 297 79105 2

Printed in Great Britain by
Butler & Tanner Ltd
Frome and London

CONTENTS

ACKNOWLEDGEMENTS

I should like to say thank you to Wilma and Colin Lawson, Douglas and Marjorie Reid, Michael O'Gorman MRCVS, David Broome, Stella Harries, Malcolm Pyrah, Jenny Pitman, Stephen Smith and George Edwins for their contributions to this book.

To my son, Michael Dickerson, who did the technical drawings, because I know how very little ponies mean to you, I am doubly grateful for your efforts. As to the lighter illustrations, we were helped in our search for budding artists by *PONY* magazine, its editor Julia Goodwin, assistant editor Diana Wallace, and their art editor Mark Gibbs, who invited children from their large and multi-talented readership to submit drawings. To each and everyone of you who did, my sincere thanks, plus a promise that they will all be treasured and kept for, hopefully, another time, another story.

There are many others who may be surprised to know how much they have contributed to the following pages, but I should like to express my gratitude for the long hours we've spent in pleasurable pony talk and for helping us so many times in so many ways. They include Chris and Lorraine Drewe, Jill Gould, Lynne Morgan, Valerie Hosegood, Marguerite Claxton, Pat and Don Norton, Sheila Roberts, Irene Black, Philip Tuckwell, Joanne and Angus Gilmour, Vi and Vicky Jackson, Patsy Miles, Sarah Champness, Peter Lamb and Sally and Brian Knight.

1 THE TRAP

Pauline Sim (age 15)

Many moons ago, when I was a small, plump Scottish girl child, I would gladly have given my parents away in exchange for a pony. They simply could not understand my addiction to such a dangerous and costly pastime, so my dreams remained unfulfilled. My stage-struck mother, meanwhile, wasted fortunes on elocution lessons and having my teeth fixed.

Based on the natural assumption that the greatest gifts I could offer my own children were surely riding lessons and a pony, I worked hard to redress the balance from the day they were born. And so it came to pass, that when he was seven, my son was kicked by a Shetland and has had little time for ponies or his mother ever since, whilst his sister obviously felt obliged to reward their mother's dedication.

I have often wondered if, in years to come, Sam will make all the same mistakes with her own children. Will she be rushing out to purchase a Falabella to share the cot rather than a first cuddly toy? Will she lessen the odds against her children not being keen to ride by teaching Hor-sy instead of Mummy and Daddy when they are learning to talk? Will they be photographed sitting in a saddle with apprehensive looks on their faces, as mine were before they could stand on their own two feet? Will they be able to spell equestrian before they can spell cat? How can you fail to convey to your children your enthusiasm for something which has featured greatly in your own childhood? I wish I knew. I should like to know if Sam would have discovered the joy of ponies without my prompting.

Some children become what they are in adulthood despite their upbringing, but for others it's hard to believe that a world exists outside of their parents' work, sport or hobby which governs the way of life. If your father is Harvey Smith will you become another Rudolf Nureyev?

Papi

We were caught in a tea shop in Woodbridge. I have always been drawn to livestock columns the way most folk are drawn to horoscopes. Today, according to our local newspaper, a breeder of part-bred Arabs was selling out. Her youngsters were available at give-away prices. Tim, my husband, was performing a juggling act with a baby, a cup of coffee, and a forkful of sponge cake for our four-year-old son.

The conversation, which was to have such a serious effect on our lives, went something like this:

'What are you reading?'

'The Ponies For Sale column.'

'Do we need a pony?'

'Well we will when they're older.'

'Isn't it a little impractical for as long as we are 5,000 miles away in the Middle East for most of the year?'

'They're like vintage wine, you buy them young, and they improve with age.'

We were, after all, on holiday with nothing much planned for that day. We could have a pleasant run along the coast to see this lady and her ponies. We wouldn't have to commit ourselves to anything – just be a part of the band of genuine time wasters who reply to advertisements for a hobby....

He was a yearling called Sutherland Papillon and my heart soared the minute I set eyes on him. He was like warm velvet to touch and simply floated as he was run out for inspection. I didn't really know what I was meant to be looking for but whatever it was I was sure this pony had it in plenty.

'Take your time to think about it,' said his wise breeder.

We took all of thirty seconds. On a beautiful summer day in 1975 we had got our first pony. We were so naive, we would have paid the money and made off with him there and then, but it was this lovely honest woman who insisted we had him vetted before she accepted the cheque.

I called our local veterinary practice, who said that Mr Henderson was their horse specialist and he would call out there the next day. It was worse than waiting for a first baby.

Next evening he called to say there was a slight problem. He was almost certain it would be all right, but he wanted to check with the vet who had castrated the pony, as he could feel a slight lump and wondered if one testicle remained after the recent gelding.

'We just might have a rig here,' he said.

'Is that serious?' I asked, agog at this new development.

'Well I shouldn't recommend it in a child's pony,' he went on. 'Mark you, having said that, I had one myself when I was young, and it turned out to be an absolute jewel.'

'Of course,' I flustered, 'Well.... and good heavens ... you're perfectly developed, and you've even fathered three children haven't you?'

'I meant,' he said slowly and patiently, 'I had one *PONY* with the same problem when I was young.'

Twelve years after the purchase of Papi, I am to the World of Ponies what Lady Godiva was to three-day eventing: an enthusiastic amateur with good intentions. It is not of course advisable to buy first and worry later.

I have learned much over the years, thanks to the generosity of friends and various experts who have taken the time and trouble to teach me, but I still have many miles to go.

I should not be what I am today, an unpaid groom and nervous wreck, but for the boundless enthusiasm of my daughter Sam and her Fairy Godmother Wilma, who has provided and produced most of the ponies you will meet in the following pages. It is with their help that I have compiled a Novice Working Mother's Guide to steer others who are drawn irrevocably into The Pony Trap.

Gay

A problem which faces a great many pony parents who don't ride themselves, is this. If they buy a pony for a very young child, who will take it out on rides? Not many people are in the privileged position of having endless acres of their own parkland to ride over and, except perhaps in the more remote areas of Britain, there aren't all that many safe rides away from busy traffic spots. We had to deal with this situation when Sam and Papi were seven years old and ready to get their act together.

I know there are many show ponies who never set foot on a road just in case they hurt themselves, or are simply unsafe in the hands of young passengers, but keeping a pony going round in circles on a lunge or plodding around the same boring paddock was never our intention if we were to enjoy riding.

It became perfectly obvious that we had to look out for a nanny for Papi, and that meant a return to the saddle for me. In the meantime, a new friend and neighbour, Lorraine Drewe, was taking Sam out on hacks around our village, which usually turns into a mini Silverstone at weekends when the motorbike riders come out in force.

One day when Lorraine arrived to pick up Sam, she was sitting on a very pretty New Forest bay mare that I had been admiring for a week or so in their yard, assuming it was one of the liveries they keep there. As fate would have it, she was for sale, and came highly recommended by Lorraine as a babysitter for Papi.

One morning I tried her out on a ride in a field with Sam, who was illustrating at one point how Lorraine had taught her to jump straw bales. A combine harvester was churning away in the background, but it bothered Papi not a jot. However, as is often the way with ponies, something small, harmless, but unexpected – as in this case a miniscule field mouse – fled from under the jump as they went over. Papi in fright swerved and took off like a train, reins and rider hanging round his neck as he fled from the field towards a track leading to the open road. Fortunately, Gay and I were positioned near enough to cut the corner and get to the track before our Papi. I positioned her across it and waited to see whether the small pony would collide with the big one or swerve and make off again at top speed before I could grab a rein. I now know how a picador feels as the bull charges down on him. I was convinced that the collision, when it came, must have knocked a permanent dent in Gay's belly, but as we gathered up our two small charges she just stood there, calm and collected. Her Nanny's audition passed with flying colours. We had found Mary Poppins. She actually began life fifteen years ago as Niton Forest Fern, but apparently was given the stable name of Gay as a foal. It wouldn't perhaps have been my immediate choice for her, but she knew her name and she was happy with it, which seemed the important thing.

Gay's a good sort, with the conformation for weight carrying, and placid or idle in disposition, depending on how charitable we are feeling towards her at the time. She is also comparatively cheap to run, doing more miles to the gallon than any pony I have ever known. In the stable she is fed less than our other two ponies, Dusty and Abbey, although she is over 14 hands and they are 13 and under, and she still gets fat.

The first year we let her winter out in a large field where she could go round the grass like a high-speed lawnmower, but one morning we watched her potter towards us displaying the first dreaded signs of laminitis (see page 66). We walked her the mile home and put her on a near starvation diet of bran mashes with Epsoms and the minimum of hay, just enough to avoid upsetting her kidneys by too drastic a reduction. We then returned to our task of chipping ice off the trough back at the field – only Gay could get her first and only attack of laminitis with snow on the ground in the first week of January! The danger signs passed within a couple of days, but she is still a terrible glutton and has to be watched like a hawk in a field when she's fed with the other ponies and goes over the left-over hay like a hoover.

And as I suffer from vertigo on anything over 14.2, especially after years of comparatively small Arab horses in the Arabian Gulf, it is an incentive therefore for Gay and her rider to serve a life term of Weight Watchers together. It also means she has built up quite a fan club of young children willing to take her out for me, being one of a very rare species – the genuine family pony.

Abbey

The story of Abbey's birth starts like a fairy tale by the Brothers Grimm. Once upon a time as cold winds blew and snow lay on the ground around Lindores Abbey in Scotland, a fairly elderly Arab mare was, to everyone's surprise, safely delivered of a daughter, who was wrapped in a blanket, taken into a warm stable and christened with the name of her birthplace. Wilma had suspected the mare was in foal, and had been keeping an eye on her. It is unusual to say the least for mares to give birth at such bleak times of the year, but Abbey is a pony full of surprises. Even her mother, who died not long after she was born, would have hesitated to call her beautiful, and she was, as we say in Scotland, born 'on the wrong side of the blanket', which is to say, out of wedlock. A dashing Welsh Palamino stallion, renowned for his jumping ability, leaped over a fence one day and led her pure white mother astray. Cut off from her rightful inheritance, Abbey's future remained uncertain, until as time passed she began to look less like a cream bun on stilts and more like a promising working hunter pony.

This idea was born one day when Abbey was four and had just been introduced to a saddle. Sam, aged nine, was trying her out for size in the fields at Lindores. The clock struck four and the buckets signalling teatime started to clank away in the yard. Abbey stood still and listened then set off at a trot and then, despite the pleas from her young rider, into a canter towards a five-bar gate.

Wilma is wonderfully calm in such situations. 'Sit on Sam, she'll stop at the gate,' she called without a flicker of concern in her voice. Seconds later, we all watched open mouthed as this promising combination sailed over the gate and into Abbey's stable and the waiting bucket.

Abbey has remained bold, honest and anxious to please. The bonus for a child is that she is a very cuddly pony. Having been an orphan and a social outcast early on, she now grabs all the love and attention she can get. She has given Sam a lot of confidence as she is a good all-rounder and tries hard. Whoever gets her next will be very lucky.

Dusty

I wish we knew more about Dusty's background. It is obvious he started in good hands and went to the right schools. He is a proper gentleman with manners to spare. Yet clearly, later on in life he had been badly brought up by foster parents. The British Show Pony Society helped me trace previous owners in vain. People move on with the years, so his occasionally wayward behaviour remains unaccountable.

He was recommended to us as a schoolmaster for Sam a few years ago. He had had a seemingly charmed life, judging from his track record at many top-class showing events,and his list of successes accompanied a photograph of him looking gorgeous. Even at the age of fourteen, he was clearly a pony who knew the job and was worth considering.

Viewing was at a dealer's yard in Wales, and on the journey through December sleet and snow I warned Sam that he wouldn't look young and dappled any more, and that the best he could do would be to teach her ringcraft and see her safely through Pony Club events. His days of top-class competitions were most likely behind him. But we know some geriatric ponies who are still looking great and going strong, so I think she was reasonably optimistic.

We were stunned at the first sight of him. He looked more like a sheep than a pony, with a thick, dirty, cream-coloured curly coat and eyes sticking out like organ stops on a small bony head. As we approached him his ears went back and he became threatening and clearly apprehensive. Sam produced the Polos and carrots we had brought for him and it was like magic, he fell upon them and we made friends.

I was frankly appalled at his condition and amazed at how such a pony who had obviously put in many years of first-class service could be allowed to deteriorate like that, which is a totally naive attitude of course – there are hundreds of Dustys. There is only one way for a top show pony to go and that is down. Many of them can't even be recommended as a safe ride for a small child to hack out on because they have been used to skilled older riders who can get a better tune out of them. There was every likelihood that from under the sheepskin clothing a wicked wolf would emerge. Certainly his appearance did nothing to commend him. Yet for all that, he held himself with great pride. Dilapidated and with all the odds against him, he didn't look like giving in.

We took him on trial. Wilma came down to see him and it was obvious straight away she thought the money being asked was ridiculous. 'We give poor souls like this away to good homes in Scotland,' she exclaimed. Sam, however, never wavered in her love for him, despite the fact that he was carting her daily round the paddock at break-neck speed due to having a mouth like a plank, and a firm belief that you had to go faster and faster in order to survive.

'Good God,' said the vet when he first saw him, 'Where did you pick this up.' Our hearts sank as he looked him over. 'His front legs come out of the same hole, he's riddled with worms I'll bet you, you'll never get food into him until we do something drastic about the teeth.' He was falling on his food, but 'quidding' – dropping it in lumps out of his mouth because of an inability to chew properly. 'Can we get a jockey and see what else is wrong with him?' We explained that the jockey was having trouble finding the brakes and steering, and if we wanted to finish the vetting before nightfall we had better test him on a lunge.

Dusty passed his medical but as the vet drove off he asked, 'How much are they asking for him?' I was frightened to tell him. I expected him to double up with laughter as everyone else had done, or to certify *me*, rather than the pony. 'That's nothing,' he said, 'he's a real little pro. By rights he shouldn't be standing up in that condition, he could come right in the body, it's what's in his head you have to worry about.'

It was decision time, he had passed the vetting, he was quieter on roads than in enclosed spaces, and in fact hacking seemed to relax him. We had had him nearly a week and the next day, Christmas Eve, he was due to return. I tried not to be swayed by the craftily positioned stocking bulging with Polos, a new headcollar, apples and a Thelwell card on top, which said, 'You may not find the perfect pony straight away, but sooner or later he'll find you!' How did we come to be involved with this dubious wreck with a heart made of sunshine? There was no

8

question of him going back. I sold a piece of jewellery that hadn't given me the pleasure in years that he'd given us in a few days.

Our purchasing of Dusty is the perfect example of how not to buy a child's pony. I have to say his progress back to being a happy, fit and well-behaved animal was an uphill struggle with many setbacks along the way. I should like to report I am sorry I ever bought him. The fact is, I do not regret one second of the five years he has been with us. He has taught us more about patience and understanding with ponies than a newer model straight off the production line could have done. Today, he owes us nothing and we will never part with him. Compared to some of the misery and suffering which ponies endure at the hands of self-styled animal lovers, Dusty has had a good life.

2 OBTAINING THE PERFECT MOUNT

Ruth Kearney (age 18)

A bad pony is as expensive to keep as a good one. A pony with a kind temperament and good manners, who is a willing and safe ride, has got to be worth a lot. You are paying for someone else's time, patience and expertise spent in breaking, schooling and producing him to a high standard. Once you have bought him, provided you take care to maintain his good qualities, you shouldn't have any difficulty in finding another home for him, should you need to.

The idea of a pony who may win rosettes, but who is unsafe in a child's hands outside a ring, or who is a fiend in the stable, has never appealed to me. You must of course decide what is important to you. If it is a row of cups along the mantelpiece, you may have to settle for a pony who is not best suited to hacking. If you want a genuine bomb-proof child's first pony, you will indeed be lucky to find one who also has the presence and appearance to win a First Ridden class at a show.

Although type, condition, training and a showing record most certainly have a bearing on the price, generally speaking a pony is at his most valuable at six years old, provided he is mature and well trained. From the time he is eight or nine he is already on the decline, and by fourteen or fifteen he is beginning to age, so beware of buying a pony over the age of ten as an investment, thinking you will have a couple of years of success at shows and money back at the end of it. On the other hand, a schoolmaster pony of advancing years, having settled into his work, can be invaluable for teaching a learner rider.

Never buy on impulse, take time to do your homework and to fully understand the moral and legal commitments involved when you take on an animal. For instance, a warranted pony sounds like a simple, money-back guaranteed proposition, and for sure a warranted pony may be preferable to one that is not. In theory the pony is backed by a statement of facts supplied by the seller, but only written ones may be produced as evidence should the need arise later on. If they turn out to be untrue, e.g. after a vet's inspection, and it is confirmed that the pony does not meet the warranty, then the buyer can claim compensation. A solicitor or maybe even a barrister will then be needed to conduct the case. If the verdict is in favour of the buyer, the court may order the seller to pay the difference between the value of the pony as it is and the value had it been as warranted. The pony remains the buyer's property. So all in all, unless it is a top-class show pony and has changed hands for thousands of pounds, there may not be much joy in this costly, time-consuming and very depressing process.

By Word of Mouth

This is the ideal way to find a pony. Go and see it in work, see it ridden by both its present and future rider and, if satisfied that it is the one you want, ask if you may have it on a trial period. This is obviously in the buyer's interest, but not necessarily in the seller's, so don't immediately suspect that the owners have something to hide if the request is refused. Some owners will be happy to send an animal they know will live up to all their claims, others will take the view that he might be at risk from bad handling, indeed injury during the trial period, when he might be unsettled by the move in any case.

Another advantage of buying through friends or friends of friends, is that an arrangement can usually be reached quite amicably, but however friendly an agreement, never part with your money until your vet, not the seller's, has given the pony a clean bill of health. Find a horse and pony specialist rather than one who operates a small animal surgery where he is more at home filing budgies' beaks. He will vet the pony and issue a certificate accordingly, having examined heart and lungs, and checked the age etc. Tell him what sort of work you will be asking the pony to do, and if it is for showing, ask him about having it measured (see page 122) to make sure its height is as stated. Don't be afraid to reveal your lack of experience, if that is the case. Vets are like doctors, nothing surprises them and they are always glad to advise on suitable diets and other welfare issues.

Sales

There are bargains to be had at sales, especially those organised by pony societies to assist breeders in selling their surplus stock. If you consider yourself an expert and feel confident you know what you are doing, you might pick up a genuine treasure, otherwise it is a bit like the second-hand car market, all the bad points will be covered up and all the good points grossly exaggerated. On browsing through the catalogue, you will marvel at how they managed to assemble the absolute cream of this year's equine collection all at one venue.

A rider in our village bought a mare from a couple who refused to refund the money when the horse took to rearing, with its rider aloft, every time it was asked to leave the stable yard. The previous owners were flabbergasted at this news. The horse had of course been a paragon of virtue when they had it. After fruitless legal battles, the new owner decided to cut her losses and put the mare into a sale. In the intervening weeks the horse was sent for schooling. It continued to rear, even with an experienced horseman on its back, but when fattened up like at goose at Christmas as she was led into the ring at Stow Fair, she looked superb. We all heard the shouts of joy from the owner as the mare made the top price of the day, and the sins of the previous owners were transferred to the next.

As the crowds dispersed, a vicar's wife who had just sold the family pet for meat prices, now that her children were more interested in discos, was heard to comment, 'Well, when it comes to horse flesh, none of us is entirely honest, are we?' Nobody took up the challenge.

Dealers

Next to horse sales, buying through shady dealers is probably the next best way of ending up with a pig in a poke. Of course there are honest dealers who handle high-class animals for reputable owners. There are also the ones who will buy in any pony if they can get it at a low enough price. It might then be well schooled, until it is thoroughly brainwashed, shampooed and set, and advertised as the perfect child's pony at three times the price the dealer paid. The danger to the unsuspecting buyer here, is that as soon as it is in its new home where it has an easier time – not given a bash in the teeth every time it misbehaves, or worked off the legs every day until it is too tired to argue – a very different creature from the docile chap you first fell in love with might emerge.

In a good dealer's yard, you should be able to get an assurance that the pony can be returned if it does not live up to its trade description. An honest dealer has his reputation to protect.

Pony societies and horse agencies

Pony societies, who deal with pure-bred native ponies, give helpful guidance to prospective buyers. The National Pony Society at Brook House, 25 High Street, Alton, Hants GU34 1AW, has a list of the various breed societies and the sales secretaries.

Horse agencies do much of their business on the telephone, finding out your requirements and offering you a list of possibilities to visit within, say, a forty- to fifty-mile radius of your home. The commission will be around five per cent and agents' addresses can often be found in horse and pony magazines.

Saddlers and horse feed shops often carry private advertisements with photographs of ponies for sale or lease, and contacting your local branch of the Pony Club often produces a genuine all-rounder. Pony Club instructresses have a chance to assess individual ponies at rallies and camp. It is worth tracking them down to ask their opinion. The trouble with taking on somebody else's Pony Club pony is that if your child joins herself, she may forever be compared to the previous rider, never feeling that the pony is truly her own.

Always beware of buying a pony which has had a good rider. This is not to say that if the child who shows it to you rides well, don't touch it, but that after an experienced, sensitive rider, who has probably owned the pony for some time and can make it go like a clockwork mouse, once the pony is in the hands of your smaller/younger/more nervous child, it can undergo a dramatic change of personality.

But it is not all doom and gloom. Just because I do not know anyone who tells the whole truth about a pony they are selling, it does not mean that they do not exist. Also, there are many laughs and surprises along the way, especially if you buy through the published advertisements. The local papers tend to deal with family ponies, but *Horse and Hound* is the *Exchange and Mart* of the more competitive animal and a weekly read of what it has to offer for sale is better than anything Mills and Boon has ever come up with. Here is a quick translation of some of the more common descriptions:

GENUINE REASON FOR SALE: We need the money to buy a better one.
SUITABLE AS A CHILD'S SECOND PONY: Lethal as a child's first pony (mouth like a plank, bucks, rears, naps etc.).
KEEN PONY: Bolts the second you touch his back.
NOT A NOVICE RIDE: Only for children who have obtained their Fellowship of the British Horse Society.
SHOULD TAKE THE RIGHT CHILD TO THE TOP: the top of the nearest motorway, ten-acre field, tree, queue at the casualty department.
BOMB-PROOF: Pure-bred Arab pony raised in Beirut.

HAPPA

In 1986, the Horses and Ponies Protection Association assumed responsibility for over 600 horses, ponies and donkeys at their Greenbank and Shores Hey Farms, or within homes where they were given on loan to knowledgeable people to care for them. Selected HAPPA homes are monitored on a regular basis by local inspectors, and there is usually a waiting list for those animals, including children's ponies, which can be ridden.

It is worth getting in touch with them if you feel you have the necessary experience and a suitable home to offer a pony who has been rescued by this charity. There are many more horses, ponies and donkeys whose riding days are over, who could be seeking 'redundant mother with similar problem', and no doubt suffering withdrawal symptoms. Their address is Greenbank Farm, Fence, Burnley, Lancashire (Tel. 0282 65909).

To loan

The old adage 'neither a borrower nor a lender be' is probably good advice for the novice pony person. The worry that is attached to owning a pony is doubled I reckon when that pony doesn't belong to you. Any day now we shall have to let Dusty go on loan, or have Sam's legs surgically shortened, and it is not a decision I'm looking forward to. To sell him of course would be unthinkable, like selling our grandfather. When he goes I shall probably spend sleepless nights missing his daft ways and wondering if he is happy.

Sadly, even the most loved ponies are outgrown all too soon and have to find another jockey. Teenagers come under increasing pressure at school to concentrate on exams, old ponies need to be kept in regular work to stop them rusting over, and young ones often have to shift to more experienced hands for a while. Mothers go through bankruptcy, pregnancy or lunacy finding that they have collected too many animals to be able to care for them all properly. Whatever the reasons, loaning can be a dodgy business and shouldn't be gone into on impulse. But it must be said that many people give or take ponies on loan and find the arrangement more satisfactory than buying or selling.

It is ideal if you can lend to a friend or at least to someone who has been recommended by word of mouth where the home is close enough to keep an eye on the pony in case unforeseen difficulties arise. Your local branch of the Pony Club can usually help here. They will quite possibly keep a 'book' of ponies in need of permanent or just temporary accommodation. It is vital in any case to check out the home and surrogate parent and rider to convince yourself that your cherished animal will be in good hands. It is unlikely that you will feel he will be better off than if he stayed with you, but it is possible, and the way to find out without a total commitment is to suggest a trial period so that you can both get out of the deal if there is a change of heart.

Hard though it may seem, it is best to resist the temptation to drop in to see him too often, although you should make it clear in a legal document that you reserve the right to visit as often as you wish. Unwarranted interference on the part of the owner could very well turn the arrangement sour and give the borrowers every reason to dump the animal back on your doorstep. In any case you should already have inspected the premises and convinced yourself before he goes that his foster family has enough knowledge and concern for his welfare to get in touch with you should the need arise.

You must also state in a document that you can remove the animal if he is not being kept to the agreed standard. Shaking hands on a deal is not enough, you must enter into a business agreement of terms. If you merely want a good home and you are not asking money for his use you are obviously in a better position to recover the pony at short notice and stipulate his duties, but you should still state in a signed document that you have the right to remove the pony if not kept as agreed. It is also important to make clear that he will be ridden by only one, named rider.

You must, of course, and first of all, lay down in writing that the borrower shall assume complete financial responsibility in addition to caring for the said pony for a stated period. This must include the insurance. When you lease a pony and even a nominal amount changes hands in a legal agreement, you could find it well-nigh impossible to take him back on the spur of the moment. As soon as you accept money for his use, you immediately reduce your rights to the pony.

You may have stipulated that he is to be used only for hacking and Pony Club activities, but find that his every weekend is spent whipping round show jumping classes at as many shows as the borrower can fit in. A legal document prepared by your solicitor on your instructions, to be reviewed every six months, is drawn up to protect your interests and the borrower's. This ensures that you don't take the pony back for personal reasons, only if the borrower does not comply with the terms.

However, you may still have difficulty enforcing them. I know of at least two situations where the owner has fared considerably worse than the borrower, to say nothing of the hardship endured by the pony from what was originally intended to be for his good. In the first, due to an owner's riding accident and resultant injury, she loaned her pony to a reputable riding school, and got him back four months later, physically sound, yet totally lacking in interest in his work and far less sensitive to his rider's aids – which will not come as a great surprise to those who feel a riding school is a last resort.

In another case, a small pony was loaned to a supposedly knowledgeable home so that it would not be turned out, kept short of work and run the risk of laminitis. The animal, however, contracted the disease within a few weeks of the transfer. Rather than admit it, the borrower furtively tried to sell the pony at Reading Sales at meat prices rather than let its original owner find out the worst. By the time the full horror story came to light and the frantic owner found out where it had gone, the pony had gone to that great collecting ring in the sky, via a brief spell at a riding school in Norfolk. The borrower still claims she had only the welfare of all in mind when she behaved in this totally incomprehensible fashion.

We all know too many sad tales of once honest ponies who are leased as show jumpers, and, soured by the demands to jump higher and faster, are returned to the owner in a sorry state. It must also be said that there are too many owners who are prepared to take this chance for a monthly pay cheque and the possibility that if things go well the pony will be worth even more to them at the end of the term of loaning. There are two sides to every story. A borrower can find that the pony is not at all as described and quite unsuited to the job he was meant to do.

Where then does it leave us? It leaves me I'm afraid with the prospect of the coming years spent wrapped in rugs with a floral hat on my head, and Dusty pulling a trap for our daily jaunt to the village post office or to drop in on friends in rural localities for a cup of coffee and a carrot.

Winifred Hoyles

The Ideal Pony

So how do you recognise the perfect pony when your horse sense as well as your bank balance is limited? In-the-know Pony Mothers wax lyrical about a pony's conformation, movement and presence, often ignoring his usefulness. They will tell his head's set on right, his neck's upside down, his quarters dip down or his belly tucks up, when all you can see is a nice little soul that surely any child would be thrilled to call their own. The experienced Pony Mother's X-ray eyes will have noted the warts, splints, cracked heels, wall eyes, capped hocks and wind galls, before you have even worked out whether he is a brown, bay or liver chestnut. It will take you years to catch up, but because the conversation rarely strays from the subject of ponies when parents get together, you will learn a lot by watching, listening and reading.

However, so much has been written about a pony's conformation that is well beyond the grasp of the amateur. It can be very confusing if you haven't studied many examples of good and weak points under the guidance of an expert. The reason why a pony should have good conformation is not just concerned with his appearance, but with the work he has to do and his general health and condition. A pony with a good shape, which is basically what conformation means, will attain and retain condition more easily. He will be less prone to problems with his heart and lungs, or to suffer stress-related injury to his back or legs when carrying a weight. Weak points, therefore, can lead to unsoundness. With every pony you see, build an imaginary frame around it.

A well-built pony should fill a square with each part of his body being in proportion to the rest. Relate the bad points to potential problems. If he is too narrow in the chest, for instance, his breathing could be restricted. Too-long backs are undesirable where he needs strength for galloping or jumping. The shoulder should be gently sloping to allow

for plenty of muscle rather than straight, and the neck should form a slight arch from the poll to a clearly defined but not over-pronounced wither, which should never be lower than the quarters if you study the top line as a judge does. In between, behind the wither where the saddle sits, a gentle curve is desirable for a comfortable ride, not one with a great dip as you tend to see more in older ponies, or a 'roach' or flat back, which denotes inflexibility and can cause stress on the vertebrae. Standing behind the pony, his hindlegs should be a matching pair and his quarters be strong, rounded and level.

Back to the square, where in the perfect pony the hock should be within a straight line from the end of the quarters to the back of the pastern. Very bent hocks also tend to be weak. The forearm should be strong and straight and the knees flat in front – beware of knobbles which are indications of previous injury or strain, and are especially unwelcome in young ponies.

As to the head, you will hear experts say that if it's all right the rest will follow. Don't take that literally, but you will soon spot the difference between a neat, well-defined head, in proportion to the rest of the body, as opposed to one which is overlarge and common. Eyes do a lot to enhance his appearance. They should be big and kind with plenty of forehead in between. Small pony ears are another sign of quality.

His feet should be round and well shaped, not boxy and upright like a donkey's. Beware tide marks on the feet which indicate he has had laminitis. A well-set-on tail is neither too high as in Arab ponies nor too low, which occurs more in thoroughbred types. A pony with nice movement i.e. which flows freely from the shoulder in a nice long stride and not bouncing from the knees, will give a smoother ride and be less likely to stumble.

They say that a good pony is rarely a rotten colour, and in general this seems to be true. There is a fair bit of colour prejudice among judges. Popular showing colours are bays, liver chestnuts, blacks and greys. Chestnuts, palominos, duns and roans do not stand out so well on appearance alone and it is rare to see a piebald, skewbald or spotted pony highly placed in ridden pony classes.

Last, but by no means least, the ideal pony has that all-important presence, the look that says 'Who's a pretty boy then?' – an indefinable quality of communication which sorts out the stars from the chorus line. He should look alert and intelligent, and be of a kind temperament which makes him an honest ride. Remember, a pony you buy is supposed above all to be suitable for, and to be ridden by, a child of the appropriate age group – a fact often overlooked by parents and judges alike. Good conformation, good presentation by rider and producer to play down his faults and enhance his good points, coupled with the essentially good temperament of the child's pony, are the things that lead to show-ring success.

Although it is important for a child to find a pony that she cares for quite passionately, I don't think the decision should be left entirely to them if they are very young. On the other hand, so many ponies are the mother's and not the child's choice, which is equally unfortunate as then the care and exercise become all duty and no pleasure. A beautifully mannered, perfectly proportioned show animal can be a mother's delight and the biggest bore in the world for the jockey who would prefer something she can have fun with, even if it is on the naughty side ... which children love. Sam recently exercised another pony for a friend who was short of a rider. 'Was he nice?' I asked. 'Well he was okay, in fact he is quite nice, but he doesn't *do* anything.' Now I actually *prefer* them not to do anything, but I have to concede that

Gillian House (age 14)

ponies as well as children have to be allowed a bit of naughtiness sometimes. 'Be to his virtues ever kind ... and to his faults a little blind.'

NO VICES

This is a label often tagged on to a pony for sale, and there could well be a difference of opinion here about what constitutes a vice.

Rearing would come top of my list. Not in any circumstances would I allow a child to mount a known rearer. It can take a split second for a pony to stand on its hind legs, unseating and possibly causing serious injury to the rider. What is more, it is a trick that a pony tends to repeat once he has got away with it, and confirmed rearers are almost impossible to cure. Sending him away for schooling could prove a total waste of time. However, if he is doing it because a badly fitting saddle is hurting his back, or his mouth is sore, or he is beside himself with excitement, it may be possible to remove that cause. The professional handler will try to re-train him by dropping the hands forward when a pony rears, always keeping him on a loose rein and leaning to one side to avoid a bloody nose whilst driving him forwards. Or he might turn him in a circle to avoid a repeat performance. Some say you must not hit a rearing pony, he will only rear higher. Others will tell you that if you break a bag of water between his ears as he rises, he never repeats the habit – he thinks he's bleeding to death apparently. Try explaining that theory to a small child frightened out of his wits by a pony displaying the most frightening method of resistance out. Get rid of him.

Napping ponies can turn into rearers and also need firm handling to cure them of the habit. Unless a pony has been well schooled to move forward when asked with a leg aid, he may start by offering resistance when leaving his friends in a yard or a line of other ponies at a show. The battles which result can put children off forever. This sort of thing makes a child look as though he cannot ride, which he will hate in front of his friends, so don't buy a napper. If the pony you already own develops the habit, send him to an expert for re-schooling.

The same applies to *Bucking*. If it becomes persistent and enough to unseat the rider, have it cured quickly – nicely mannered ponies don't buck. Get help from someone who will handle the pony firmly. The theory is that he cannot buck and go forward at the same time. It might be due to high spirits caused by overfeeding or not enough exercise, or too much. He could be superfit. Dusty, on occasion, has resorted to bucking if asked to jump when the ground is too hard for his elderly legs, which is our fault for not being more considerate. When the jarring stops, so does the bucking. Always look for the cause if a pony develops a vice which is out of character.

Stable vices are less easy to detect when you buy a pony. Sometimes in a new home the very novelty of the surroundings stops the boredom which created the bad habit ... and he may be cured temporarily at any rate.

Kicking, stamping or pawing in the box ... Abbey's favourite is knocking on the door, she loves the noise, but is the only one who does as dawn breaks, so we tried various methods to cure it. The one which didn't work was turning the pony out to break the habit, but making a lining from gorse for the door was a success and soon spoiled Abbey's fun. Other methods worth a try are, especially with a young pony, providing toys such as a turnip or half a rubber tyre on a rope; turning the box into a padded cell with bales of straw (minus the twine); hanging sacks of straw or nailing rubber tyres in his special noisy bit (professionals put hobbles or kicking chains on). Rushing in at once and tying his head up tightly for an hour is effective, but it's like house training a puppy, you have to be on the alert and ready to deal with it at once. There could be a simpler remedy, like providing him with a really de luxe deep bed to encourage him to lie down more or making sure he hasn't got lice or even mice in his bed. Perhaps he isn't getting enough exercise. As a vice in a child's pony I would rate it preferable to biting.

Biting. Too often the cause is feeding titbits. I regret to say that Dusty has enough pocket money each week for one tube of sweets. He never bites anyone, but is a Polo junkie and gets them as a reward for being especially good above the call of duty.

Ponies who bite the hands that feed them are as irritating as authors who lapse into phases of don't do as I do, do as I say – but a pony who nips certainly doesn't deserve titbits.

At Wilma's all small children who bring titbits have to put them in the ponies' tea buckets, which is hard but proper. Ponies often nip when badly groomed or when having a girth tightened, so it makes sense to take extra care over grooming and to tie him up before you start. A sharp slap on the side of the muzzle is the answer if you're quick, but it must be done the same instant as some artful dodgers can nip and shoot back all at once to avoid the punishment.

Crib-biting and wind-sucking are serious in that they can lead to unsoundness. A crib-biter gets hold of a solid object, such as the top of a stable door, with his teeth and swallows air. Too much of this and the incisor teeth become damaged and consequently he cannot graze properly. Wind-suckers don't need anything to bite on, they simply arch their necks and swallow noisily. Recently a woman near us left a non-horsy person in charge of her home one day. Wandering round the yard, this chap was greatly alarmed by the sight of her wind-sucker, rushed to the phone and rang the vet to say there was a horse being sick and could he come out straight away. But it's no joke, it can cause colic and upset the digestive system which then results in loss of condition.

The common causes are boredom or not enough bulk food. Ticklish ponies can start the habit when being groomed, especially if they have unseasoned wood to chew. Treat the wood with a safe coating of anti-nibble preserve like Cribox or creosote. Maybe he needs a salt or mineral lick. Don't let his friends see him or they may well be impressed and take up the habit themselves. Although it is difficult to cure you can at least take the fun out of it by fitting a special strap behind his ears, or a 'flute' bit attached to his headcollar when he's in the stable. But as Lucy Rees points out in her excellent book on why horses behave the way they do, *In The Horse's Mind*, restricting neurotic behaviour is like pulling out the nails of a nail-biter, it doesn't stop the desire or deal with the cause.

Weaving is an involuntary restlessness, and if a pony can't get enough rest his general condition will suffer. Also he can go lame through the constant rocking and swaying of his head and forelegs. The best remedy is to turn him out. If you must stable him give him an extra-deep bed to encourage him to lie down, and attach a V-shaped grid to his stable door to restrict his movement.

A pony which *eats its bed, or worse still his droppings* could have a deficiency in his diet. Giving more hay and other bulk foods can satisfy pangs of hunger if he likes nocturnal bedding feasts, and making the straw as unappetising as possible will also help. Use only wheat straw for bedding, spraying it with a solution of Jeyes Fluid, or change to wood shavings or peat. Provide a salt or mineral lick, and make sure he's regularly wormed. As a last resort, muzzle him.

Cloth eaters. Ponies who chew and tear at their rugs may well be trying to tell you something rather than being downright ungrateful. If you have ever had something biting you inside your clothing, you'll understand, so check his coat for bites or lice. Check his diet too – fine-coated ponies on too rich a diet often have problems with skin irritations or vitamin or mineral deficiency – try a salt or mineral lick in the stable. If it is just habit you might break it by spraying the coat with an unappetising insect repellant or disinfectant.

Halter pulling. Ponies may first develop this habit through fear or anxiety if they're left alone for long periods. They soon discover the joy of freedom it brings and take it up as a regular hobby. Not leaving him alone is an obvious remedy but it's highly inconvenient not to be able to leave a pony tied, at least for a short while. In the case of an older pony he might be persuaded to change his mind with a short sharp thrust from a broom head up the rear and the command 'no' at the same time.

Ponies who behave badly in their boxes or elsewhere and give children a hard time, for instance when a bridle is being fitted, quite often do so because they have been accustomed to a jab in the mouth or a poke in the eye, ear or whatever. After you have re-schooled the child, the pony can be cured of the habit by not having the reins taken over his head, but having them unbuckled and passed up either side of the neck and rebuckled at the withers. Now you can move the buckle bit up behind his head and firmly round his neck, at which point he will surrender if he's at all smart. If he isn't, open his mouth and try holding the bit in the left hand, pressing it against his teeth while you slip your left thumb into his mouth to tickle his tongue. This generally persuades him to open up without further fuss, but needs patience, practice and help when tiny hands are trying to cope.

Unauthorised grass eaters. I know this doesn't constitute a vice, but it is a bad habit in a great many ponies who get the better of their small riders by stopping to eat grass instead of going forward when asked. Basically it's bad schooling, but even a nicely mannered pony can turn easily to this sort of behaviour once he thinks he can get away with it. An effective deterrent is to fix a pair of grass reins from the pony's mouth to the D rings of his saddle, preventing him putting his head all the way down to the ground. A makeshift pair can be rustled up in minutes from a length of baler twine.

3 BED AND BOARD

Nicola Garnham (age 12)

Stables

The ideal way to keep a pony is at your home, with a proper stable at his disposal. This immediately offers him greater protection from fire and theft than if he's isolated miles away, as well as providing you with many other obvious advantages. Apart from the luxury of not having to go far on wintry mornings to attend to him, or not having to begin the mammoth task of cleaning him up on the morning of a show, you can bring him in if he looks a bit off-colour, keep him out of the daytime heat and flies, and control his grass consumption in summer and his thick shaggy coat in winter.

Home stabling is still cheaper than the cheapest livery. Whoever is responsible for building yours – the local tradesman, or one of the many specialists in sectional stabling who advertise in the pony magazines – you must start by approaching your local council authority for permission. They will need to see plans outlining the proposed structure and its dimensions, and they'll send an inspector along to ensure that it complies with building regulations and that it isn't yet another eyesore, blotting the landscape. You are charged for planning permission and building regulation applications. Do you really need them? Yes you do, because if the council later finds out that you went ahead without them, you can be made to take the building down. If you have a neighbour who is likely to object to the siting of the building, the accompanying noise, or a muck heap with its attendant flies and distinctive pong wafting across their back garden, try having a friendly chat with them.

The foundations must be sound, with provision for efficient drainage, and the roof should be safe. The building must be fire resistant and have an adequate outlet for rainwater. You may have a suitable outbuilding for conversion, but it must eventually have the minimum requirements which a builder or a specialist kit manufacturer would supply. That is to say, a $10' \times 10'$ living room for small ponies, and $12' \times 10'$ for anybody over 13 hands, with a four-foot divided door and at least four-foot headroom over all. The foundations can be of a non-absorbent, non-slip material such as rough concrete, provided that the floor is slightly sloping to assist drainage.

Most of the firms which supply stables expect you to provide the foundations although they will give you advice about drains and suitable materials to use. Some will also supply the plans which you need to satisfy council requirements. Shop around, ask about accessories and other services on offer. It's amazing how much prices can vary from the five-star luxury apartments to basic bed and board. The main considerations are the pony's safety and comfort, he won't mind if his meals are delivered by an automatic feeder or in a plastic washing-up bowl.

Safety means protection from protrusions and electrical fittings which he could bump into in the dark. Stables can be damp places and a combination of water and electricity can be fatal. Cables should be enclosed in conduits (specially galvanised steel channels), and lights secured high up in the roof behind heavy glass covers. Light switches and plugs must be situated on the outside of the stable, well clear of the door.

If you can choose, go for corner fittings for feeders and water buckets. If you can't, use rubber tyres in corners to hold elephant-proof buckets for food and water, not flimsy ones with handles that can get caught in his feet. Hay nets, if incorrectly tied and positioned too low, can be another hazard to stabled ponies who will put their feet through them.

Ponies who kick very soon let in draughts and daylight through the walls, to say nothing of running the risk of injury, so stout kicking boards are a wise investment. Paint the top of the lower half of the door with Cribox or creosote to deter him from eating the new premises, or fit metal chew strips if he's a confirmed biter.

Artful ponies can make simple toy puzzles out of mere top bolts. Fit additional kick bolts and padlock bolts on the bottom, and hooks to secure both sections back against the wall when open. Securely fastened tie rings and a wall mount for a salt or mineral lick should complete his furnishings nicely.

When it comes to *bedding*, there are several choices. Ponies with silly habits like pawing or nibbling round their boxes, suit some types better than others. The most easily available are straw and shavings. Straw is the more attractive in terms of disposal as garden manure. Picking out droppings from shavings is easy, but removing wet shavings is more of a problem. Shavings are of no use to anyone, although some owners bed them into muddy patches in lungeing arenas or round gates where the ground has become poached.

Straw looks nice and is comfortable for the pony, but problems can arise with new straw which tempts a pony to eat it. Mature wheat straw is the kind you should use, but even then a Gay type will find it irresistible. Hence, you may have to consider wood shavings, bought in compressed blocks and wrapped in polythene for easy storage – expensive unless you buy in bulk. See if you can find someone to share a load, which suppliers will then deliver. Because of Gay's compulsive eating problem we have to keep both straw and shavings, to use in separate boxes because they don't mix well. I haven't used peat, but I am told by those who do, that a generous layer of peat makes a good base for straw or shavings. Ponies with allergies might even need beds of shredded paper, so it is always advisable to find out about his bedding habits before you acquire him.

Whatever you use, make sure that it is deep enough for warmth and comfort. Ponies will soon compress it by walking around and lying on it. As a rough guide you should not be able to touch the floor with the prongs of a pitchfork, and for that you could use up to half a dozen bales to get you started.

Depending on how much time you have for *mucking out*, you can either keep the bed constantly on the move, turning it over, sweeping and airing the floor daily and removing droppings and wet bits constantly, or you could adopt an almost deep-litter system which will put your back out only once a year when you totally clear the whole thing, disinfect and hose the stable down, dry it out and start all over again. This way you just remove droppings and the very wet patches and top up with a clean layer of straw or shavings as necessary. This is probably the better method where children are taking on much of the stable management, but it shouldn't be an excuse for sloppiness. One pile of droppings is to be expected, two is only just excusable and three is a disgrace.

Dirty beds breed diseases such as foul-smelling thrush in the feet, and skimpy beds give inadequate protection when the pony moves around, resulting in capped hocks as he gets up and knocks himself on near-bare floors and walls.

For tools, you'll need a shovel, a pitchfork (two prongs are best for straw), a wheelbarrow, a stable skep or a laundry basket and a stout broom for sweeping tidy. Remove the water buckets before you start, and after making the beds let any dust settle before you refill them. I find mucking out is a very personal thing, rather like cooking, in that you devise your own methods of working and resent outside interference. I've seen a stable's entire contents moved into the yard by a buxom lass who came to help in the holidays and emerged like an earthquake victim when it was eventually sorted out, and a helpful weekend guest daintily picking over the Sunday morning collection in high-heeled fluffy mules and a satin nightie! So each to their own....

Good stable management requires that a pony must have *fresh water* at all times, so two buckets will be necessary to ensure that you don't risk feeding him when he is thirsty. To water him after feeding can give him colic and other digestive disorders. If he comes in from exercise and drinks straight away, it won't harm him provided he is not to be ridden again for at least several hours afterwards. Where more than one pony is stabled, individual buckets should be marked so that each pony sticks to his own, to prevent the spread of coughs and sneezes to the other inmates. Buckets containing food and water should be cleaned daily and regularly disinfected as an extra precaution against germs.

Ponies at grass

Picture the scene. A small group of ponies graze in an immaculate paddock where the posts and rails have been carefully erected inside a thick hawthorn hedge which acts as a wind-breaker. The ponies stop a while from munching the meadow-sweet grass and meander to the cool, clear, running stream to quench their thist, lingering awhile in the shade of the old oak tree.... At least that's how it is in the movies, television serials and famous paintings by Constable, Hobbs and others who have never actually had to put up with the problems of ponies living out.

Now cut to the more likely situation where a muddy, over-grazed paddock houses too many ponies for its amount of acreage. The fencing, apparently built to keep out the enemy during World War II, is of barbed-wire construction and sagging in parts where the inmates constantly lean over to reach the better grass in the field next door. Another feature of this unique type of fencing is the way in which the barbs capture the splendour of once flowing manes and tails, now displayed at intervals where itchy beasts tortured by midges or lice have enjoyed a really good rub.

Cue in the owner of the ponies, staggering under the weight of a full water bucket in each hand, with which to refill the old baby bath which is playing the part of a water trough. Her careworn face, chapped hands and grimy fingernails belie the fact that she is one of Britain's happy band of several million owner-riders, willingly giving of her limited time and money to keep her ponies in as healthy and contented a state as she possibly can. Today is a holiday which means she can wallow in such delights as picking up copious amounts of dung, pulling up the ragwort and chucking the Pepsi cans, milk bottles and oily lawn-mower clippings back over the neighbour's fence. With a bit of luck, after giving her pony a quick rub down with a mud scraper she'll be able to tack up before it gets dark and risk life and limb for an hour out in the bank holiday traffic.

Here are some advantages of keeping a pony at grass. It's the nearest he gets to his natural habitat, especially if he's in the company of other ponies. He's naturally gregarious, feels relaxed and secure with the rest of the herd, eating grass slowly and taking exercise at will – all conducive to his physical and mental wellbeing. He'll be less bored and therefore less likely to develop vices common to stabled animals. He'll acquire a natural hardiness. Continually stabled ponies become like indoor plants which are susceptible to colds, draughts and disease when suddenly placed out on the pavement. From the owner's point of view, especially where there's school or work to attend, it's cheaper and less time consuming.

The disadvantages are that he will have less supervision, he could be stolen or ill-treated by joy riders or vandals. He will be more difficult to keep clean. If he is allowed to grow his natural winter coat, he will be unable to do strenous work – he may be healthy therefore, but less fit. Winter grass alone won't sustain him. He will need additional food and inspection at least twice a day, checking that his water supply provides him with the 4-8 gallons a day that he'll drink.

It's normally reckoned that each pony will need between $1\frac{1}{2}$ and 2 acres of grazing, although some will go off the legs on that when the summer grass appears, and starve on it in winter unless supplementary feeding is given. In his favourite dunging areas the grass will become spoiled, and he will in any case avoid eating near dung and urine. As he tends to be indiscriminate both in his grazing and dunging, very little of the two acres will be left to eat. It is best to divide the area and use only half at a time, resting, cleaning and fertilising each section in turn. The ideal is for a paddock to be rested, spread with cattle dung, harrowed and rolled (always in March and then at intervals after that) to restore even grazing.

Alternatively, the do-it-yourself fanatic might consider buying a couple of heifers to clear the place, as a fellow in our village does. When they have grown fat through eating all the rough patches that the ponies don't want and splattering the ponies' grazing bits with their own particular brand of fertiliser, thereby improving the goodness in the soil, he sells them at a profit. Fortunately, cattle worms don't live in ponies and ponies' worms return the compliment by dying off inside their cattle host.

Worm damage can at worst kill, or at least lead to loss of condition and bouts of colic. A pony needs to be treated for worms regularly, between four and six weeks in summer and every two to three months in winter. When acquiring a new pony it's safest to ask your vet to do a worm count and prescribe a suitable wormer. There are many on the market in the form of powder, granules or paste which you administer with a disposable syringe. Follow the instructions to the letter.

All ponies have some worms, the best you can do is keep the amount as low as possible. Grass-kept ponies, especially in a small paddock, should have the droppings removed daily from their pasture. Once they have lain there for more than twenty-four hours it's too late to prevent the ground from becoming infected. Extreme temperatures kill worm larvae, which is why a pony can get away with less worming in winter when sharp frosts are on the ground. Unfortunately, tropical scorchers are less common in British summers and a pleasant sunny day tends just to aid the breeding process. If a pony can be stabled for twenty-four hours following worming so that the affected droppings can be collected, that's all to the good.

Last year, during their wintering-out holiday, our boys shared their field with a donkey. Having heard that donkeys carry the dreaded lung-worm, and that a lot of horse and pony folk won't have them near their pasture, I was reluctant to put them together. The owner said he would wager any money that his donkey was worm free (betting his Ass?), but I decided to put the matter to arbitration and ask the vet for advice. He agreed that not all donkeys carry lung-worm but that when they do, they can easily infect ponies .

Cattle had been grazing in the field previously, so the chances were that the area was free of parasites before we turned them out. As a precaution we wormed the ponies *and* the donkey who then grazed happily ever after.

Poisonous Plants are many and varied, and some are more prevalent in certain regions than in others. Generally speaking, Ragwort, Labur-

Carol Thorndyke (age 16)

num and Yew are among those most commonly talked about, but seemingly innocent buttercups, foxgloves and rhododendrons are dangerous too. It's worth studying. Not many people know this, but under the Weeds Act (1959) the Ministry of Agriculture can enforce owners of Ragwort-infested land to eliminate it, so you should at least be able to recognise and know how to deal with that. Ragwort is highly toxic but is rarely eaten in its growing state, which is when the yellow daisy-like flowers appear just before it goes to seed. This is the best time to remove the plant and you'll find you can pull the roots out quite easily.

However, as it dries, then it's at its most dangerous and must be carefully disposed of, preferably by burning and not just tossed away over a hedge. Absent-mindedly leaving a pile of dying Ragwort in a field can result in a pony's death due to his liver being destroyed. Ragwort poison which can also, inadvertently, be fed in hay, has a cumulative effect and an irrevocable time can pass between the weeds being eaten and the ill effects appearing. Loss of appetite and condition, constipation, a yellow discoloration of the eye, mouth and nasal membranes and eventually diarrhoea, loss of balance and collapse, are how the symptoms manifest themselves.

You can use a suitable chemical spray to clear a paddock when the plant is actively growing, and animals must be kept off it for at least the following three weeks. Where this is impractical, the only course is regular inspection of the paddock with fork and trowel in hand to keep re-growth at bay. Once symptoms appear in a pony the destruction process has begun in the liver and there is no cure, so the time spent in finding out more about it could just save a life. Write to The Ministry of Agriculture, Fisheries and Food, 3 Whitehall Place, London SW1A 2HH (Tel. 01-233 3000), and ask for their leaflet on British poisonous plants.

FIELD SHELTERS

Even hardy native ponies should be provided with some form of shelter against cold winter winds. If you can't run to a field shelter, which is the ideal, make sure that a paddock provides a wind-breaker in the form of trees or a thick hedge, or put up 3-metre panelled fencing in at least one corner.

If you can provide a field shelter and you are choosing one from scratch, take time to work out its situation and construction. The advertising pages in horsy magazines will give you an idea of what's available in an easy-to-assemble form, and the sort of prices you can expect to pay. It doesn't have to have a concrete floor, you can add straw bedding if you want to, (mucking out to keep it clean as you would in a stable, but the chances are the pony, given the choice, will choose outside for a lavatory area). Because it doesn't have foundations, you don't need to have planning permission, but it must still be strong enough not to be knocked down on to the pony, and of course sited away from the prevailing winds. A shelter obviously shouldn't have sharp edges upon which the pony might hurt himself, or too little space to move around – especially if he's in the company of other ponies where arguments can start over who gets the best bit.

Rather like the mystery of why so few trailers and horse boxes have backward-facing designs for the comfort, safety and peace of mind of the animals when travelling, you may wonder why so many ready-made field shelters are square or rectangular and not assembled from curved sections which would be more sensible. It's something to bear in mind with custom-built shelters. For instance, an oval shape would have many advantages including better protection when the wind direction changes and less chance of being cornered by bullies.

If several ponies are sharing you will have to allow the same volume of space per pony as he would have in a loose box, despite the fact that they may well choose to huddle up in one corner for warmth and companionship. Problems can arise when a new pony is introduced to a field where others have taken up residence before him. This puts him, at least temporarily, in the position of underdog, as the rest of the herd has already established squatters' rights in the area, sorting each other out as to the bully, the natural leader, the copycat, everybody's pal etc. – just like children at school. If a measure of human aid isn't given to help him establish some kind of rank with the others and make at least one good friend and ally, he could not only be totally dejected because nobody can stand him but his health could suffer from being deprived of food, shelter and even water.

From personal experience we have learned that you cannot be sure which ponies will be kind and gentle to others and which will end up as bully boys. Their personality traits can be very different under human supervision. In our paddock, Gay, in her role as nanny, has always theoretically reigned supreme over her new charges, so as head girl she could well become the aggressor. Yet she is the most mild-mannered of beasts even when interlopers are introduced.

Abbey arrived as a five year old from Scotland where she had learned self-defence the hard way, having been brought up in the company of seven big horses who regarded her as the cross they had to bear, but at least tolerated her so long as she didn't make a nuisance of herself. Dusty arrived from the equivalent of the workhouse in Wales, tired, thin, and highly nervous. Given the characters in the scene which now unfolds, if asked to place a bet you might have put money on Abbey as the outsider with more than a sporting chance to take over as leader, Dusty as her unfortunate victim and Gay as everybody's favourite at 10-1. But Dusty got in by a short head before Abbey. Sad, skinny Dusty who needed all the care in the world, found it in Gay, who has so far missed her vocation as a brood mare. After a couple of snorts and squeals at the initial sniffing session she cuddled him to her ample bosom.

A short time later enter Abbey, who is to a herd of ponies what Princess Michael is to the Royal Family. We had no worries for Abbey who had a lifetime's experience in the martial arts. She was introduced to her new family at first over the stable door and then taken out with them on rides, as is recommended by the experts, before letting them loose together in a field. We had heard all the horror stories about what happens if you just let a newcomer in, shut the gate and let him get on with it. So we put her in the paddock first to let her examine the layout and give her a psychological advantage, then we brought in Gay who, predictably, put her head down and fell upon the grass like a refugee after a hunger strike. We watched the girls graze contendedly for half an hour or so – and then we brought in Dusty.

Sam said later it was just like the Incredible Hulk on television when something made him so angry that he expanded and burst out of his shirt. Dusty, the perfect gentleman in a stable, went berserk. First he screamed with rage, rose on his hindlegs and appeared to beat his chest. At first nobody paid a blind bit of notice, which seemed to incense him further, so like a Kamikaze pilot he launched himself at Abbey, who niftily whirled round and bit him hard on the neck as his hooves whistled past her head.

Gay even stopped eating to watch in wonder as this dreadful boxing match broke out and dust, fur, a rail from the fence and finally drops of blood flew around. It was in about the eleventh round that she appointed herself referee and stopped the fight. Placing herself next to Abbey, who was in a corner, Gay nobly took the last blows upon herself as we got hold of the fiend at last and took him away.

He went round the corner to stay with Chris and Lorraine for a few days to compose himself in their livery yard and to give his house guest a chance to make herself at home. We worried in case he'd get a complex about being rejected in favour of someone new. We visited him twice a day, taking him presents and trying to explain. When he came back the three of them had a brief summit conference in the field and when talks broke down, Gay promptly took on the role of mediator, shielding Abbey from a rather more subdued Dusty, who finally saw the error of his ways and blessed peace came at last.

For a long time Gay was Abbey's shadow, running round next to her and always positioning herself between the two, even when they slept. Now, although they live in perfect harmony, when there is any bullying to be done, it's generally Abbey who is the culprit, but Dusty wears the trousers without the vulnerability of men who need the ego-boosting attentions of two women. He's the most independent in his field behaviour. He'll happily wander away from them, leave them to go on hacks or to shows, without a backward glance. The girls are at least temporarily unsettled by separation. For all Dusty's misfortunes in later years, I'm sure he must have been well handled as a youngster and allowed to develop the strong character which has seen him through all ills.

It is a shame that human interference in the lives of horses and ponies goes so often against their herd instincts; we shut them up for long periods, continually remove them from familiar territory and friends and expect them to think as we do. We spend hours making them look beautiful and outwardly healthy to win prizes or admiration from other people and yet understand so little about their mental state. It's an old cliché, but we keep coming back to it because so much of the misery and ignorance could be avoided – 'if only they could talk'.

FENCING

Common sense dictates which type of fencing is suitable for a pony enclosure and which is clearly dangerous, but as the amount of money you can afford to spend on it can often be the deciding factor, not as many pony owners as would like to, have the ideal post and rails fencing. But as with most things to do with ponies, you will get what you pay for, and this type of fencing is not only safe, strong and long lasting – and attractive to another owner should you ever sell – but you may well recover the initial outlay in the long term.

Save yourself time and trouble by buying wood which has already been treated with preserve at the sawmill or wherever it comes from. The rails should be about 4″ wide and have a flat side for convenient fixing on to posts at two-yard intervals.

When calculating the amount you'll need to order, add a few extra for corners and gate-posts. Rails shouldn't be fixed directly to gate-posts. The horizontals should be a minimum of two rails deep, and – a small point, but one which the man who put up rails for me a while back forgot – should be fixed *inside* the verticals. It may well look prettier the other way but your pony will soon discover a very simple method of letting himself out by leaning his weight on the fence.

It would be ludicrous to suppose that if we don't mention *wire fencing* it will simply go away. It is most commonly used by pony owners and comparatively cheap. Of the many types available, barbed wire is the least desirable. The safest is a heavy-weight, smooth galvanised fencing wire which won't give way under stress. The bottom strand should be at least 18″ off the ground so that the pony doesn't catch his feet in it. Too much space, we once discovered, encourages small limbo-dancing ponies to try to push it up on to their shoulders as they wriggle underneath. Check it regularly for sagging, hammering any loose posts in with a mallet. It should at least be tight. Wire strainers, which do the job for you, are advertised in horse or pony magazines.

Hedges, with the reinforcement of an inner rail, are fine so long as they are of something like hawthorn which is think and safe, and not privet which is poisonous.

Stone walls provide excellent boundaries in certain parts of the country but the danger here is that ponies can rub them loose so they need regular checking, but that goes for any type of fencing. A pony who nibbles away at bark or fencing could be trying to tell you something about his diet, perhaps he's short of bulk food or has a mineral deficiency rather than trying to eat his way out.

WATER

This is one of the most essential requirements as shortage of it will cause him all sorts of health problems, so a constant supply of fresh water is a must. The most convenient, apart from a running stream, is a piped water supply controlled automatically by a ballcock and valve to a galvanised trough. It should be sited far enough from the gate so the ground doesn't become poached, and needs to have all-round access for a quick getaway. Mount it about 6″ off the ground on blocks, and clear of falling leaves. Whatever you use, it shouldn't have sharp edges – an old bath with a plug in and filled from a garden hose is perfectly okay once taps and any other protrusions have been removed. We have used stout plastic dustbins (the spare lids are useful as skeps for dung collecting) tied with stout towing rope through their handles and secured to the posts.

A free running stream which has safe access to it, and not one where he can sink up to his knees in mud, is of course great if you can get it. Clear running streams remain that way when their bottoms are lined with gravel, not sand or mud which when stirred up are swallowed with the water as he drinks and can give him colic. Stagnant ponds therefore are totally undesirable and should be fenced off.

In winter, lag any pipes connected to your water supply and be prepared to chip the ice off trough tops, not an easy task for a child, let alone an adult when it has built up to a really deep layer overnight. We found that floating a child's plastic football helped to keep the water moving and less able to solidify – I hesitate to recommend it, knowing that some ponies will eat anything, but ours just amused themselves by playing water polo with their noses when they had nothing else to do.

A COMBINED SYSTEM OF STABLE AND PADDOCK

This works best for ponies who have a job for most of the year if you can manage it. Basically, the pony is out at night in summer, and in the morning brought in. He's then given some hay, and whatever else he needs in the way of food, he's groomed and well after his breakfast goes down he can be exercised.

A hay net in the daytime not only gives him something pleasant to keep him occupied but it stops him getting ravenous and bolting down as much rich grass as fast as he can when he gets loose again. If he's working hard, or you have very little land, he could possibly do with a small supper in a bucket before going out for the night.

In winter you reverse this procedure. The pony goes out in the daylight hours after his breakfast. You can then make his bed – or 'set it fair', as they say in the best equine circles – ready for him coming home, after school hours most likely, for grooming and exercise. He will obviously be given extra rations in the cold weather and served his final meal and a good hay net as late as possible to last him through the long winter night. Winter grass is rough and has far less goodness in it, so be generous with the hay to compensate.

Crimewatch

If your pony lives out quite a way from you, ask if you may leave your telephone number with any householders nearby or anyone who walks a dog regularly past him. Send a Christmas card from the pony with thanks.

EQUESTRIAN INSURANCE

This is expensive on the whole, although more competitive prices can be found through organisations such as The Pony Club who get special discount rates for their members with quite a few of the bigger companies. It is advisable to shop around, bearing in mind that the cheapest offer is not necessarily the best one. I have met people who find they can undercut the average premiums by as much as 50%. It all sounds wonderful until they make a claim and so many arguments ensue. In the end they get meagre compensation.

Many companies, as we found when insuring our priceless Dusty, won't entertain the idea of old ponies. Some charge hefty increases on ponies worth more than £1,000, or on those animals who do more than just hack or attend Pony Club events.

Use the services of a reputable insurance broker who will get the best deal for your particular requirements, and you won't pay anything for the work he does for you. Usually you can get a satisfactory cover for death of the pony, his permanent incapacity, tack, vet's bills, death of rider, third party liability and loss through theft or straying.

Stolen ponies

Because of the high costs of what many pony owners regard as 'the optional extras' e.g. vaccination and insurance, often the important but expensive areas are neglected until a tragedy occurs. One of the things you insure against is theft, and there has been a significant increase in the amount of stolen ponies and horses in recent years. Although more rife in certain areas in South-East England than in Scotland, for example, it's something which could happen almost anywhere at any time, and the heartbreak involved, whether it's a valuable stud animal or a well-loved family pet, doesn't bear thinking about.

You don't have to be Shergar to be at risk, dead meat prices are high enough to attract thieves. The average gates and fencing are child's play to the more professional rustlers who arm themselves with wire cutters, mallets and all sorts of mind-boggling methods to ensure the quiet and speedy disposal of animals from their homes usually in the middle of the night, and on to the Channel ferries. They can be served up on Belgian dinner tables the same day if some of the more staggering cases of profitable rustling are to be believed.

The greatest deterrent to a thief is an animal which is *freeze marked*, that is to say branded by a humane method whereby on a dark-coloured pony the hair pigment is permanently changed so the hair turns white, and on lighter-coloured animals – greys, palaminos and duns – the hair is killed off completely and will not regrow.

FarmKey, an Oxfordshire-based company which operates this scheme with full Police and British Horse Society approval, marked around 42,000 horses and ponies last year. They have a network of operators throughout Britain, and discounts are negotiable on their charges where more than one pony can be freeze marked per visit.

Although it doesn't cause pain to the animals, the freezing cold of the iron may momentarily startle the more nervous types, but the operators are experienced handlers who rarely, if ever, have problems during the process. More likely the animal won't notice he's been 'done'. Depending on his age, breed and height, it's all over in thirty-five seconds to two minutes. Apart from a possible temporary numbness, like a local anaesthetic in the dentist, there is no lasting effect other than the fact that he can now be safely identified by a number indelibly marked under his saddle, and after a couple of days when the skin has toughened he can be ridden again.

The adorable show jumping horse Ryan's Son, idolised by thousands of young riders, now sports a freeze mark and will have done a great deal to persuade his fans to follow John Whitaker's example and have their own ponies protected by the best means at their disposal.

Perhaps we can hope any day to find owners of leading show ponies follow suit. This seems to be the stumbling block. The more important aspect of keeping a pony safe from the hands of cold-hearted criminals seems still to be overshadowed by the need to keep him unblemished for the show ring. Whilst the British Show Pony Society insist that they are by no means against the freeze marking of ponies, and that points in theory shouldn't be lost if the pony is marked (for his own good after all), until judges unite in their praise for show pony owners brave enough to mark ponies, we won't be seeing many freeze-branded models at Peterborough or Stoneleigh.

It's easy to say of these owners that they are simply foolish to put winning a coveted prize above all, and certainly it would be grossly unfair to mark a pony down for purely aesthetic reasons. Yet in the fiercely competitive showing world where exhibitors are as adept in the use of cosmetics as Max Factor, of course it puts them in a quandary, and lip tattooing is a more popular alternative.

The fact is that the recovery rate of stolen freeze-marked animals is virtually 100% and many insurance companies acknowledge this by giving discount rates on premiums once a pony is registered, so some of the cost can be recovered. When he is marked, he is automatically entered on a National Register which provides full facilities for tracing him as soon as his loss is reported.

FarmKey are in constant close contact with police, abattoirs and ports throughout Britain. In addition, they offer a reward for information leading to the conviction of anyone who steals a freeze-marked animal. With such a service it's hardly surprising that the longest it has taken so far to recover a freeze-marked animal is a fortnight. And that was only because the finder who reported the abandoned animal to the police neglected to add, along with his colour, sex and size, the fact that he had a code number on his back. FarmKey is at Banbury in Oxfordshire (Tel. 0295-52544).

Feeding

For many pony mothers the thrill of taking a pony which has cost just a few hundred pounds in the raw, and turning it into a gleaming, desirable child's mount, worth three times the price, is far greater reward than all the cups and rosettes it may subsequently win.

It can be done. Let us assume you have purchased something which on the surface resembles a cross between a shaggy-hearth rug and a camel, yet hidden underneath there is a reasonable conformation and a certain *je ne sais quoi*.... Apart from a few basic grooming tools purchased quite cheaply from any saddler, elbow grease and will power, which are free to all, are the first requirements in improving the appearance of this ratbag.

However, as all the best beauty magazines emphasise, the most expensive creams and lotions in the land will have limited results if the skin isn't nourished from within, and a good grooming coupled with the right diet will ensure a coat that glows with health. Dietary supplements which help the coat particularly are boiled linseed, milk pellets, cod-liver oil and molasses. Some folk feed wonder recipes which include Guinness and eggs, but all these trimmings shouldn't be necessary if the pony is on a good balanced diet and, most important, that he is regularly wormed to keep his insides healthy.

Water is the cheapest beauty aid there is for ponies too. Have two

Lorraine Drewe, 1987

full buckets in the stable at all times, and add enough salt to the feeds or have a saltlick on hand which will encourage him to drink plenty. Where water isn't in constant supply, always water *before* feeding.

Ponies like, and do best on, a little-and-often regime, which is after all how they eat if they are given the choice in the wild. Their small stomachs are not in any case designed for a plentiful feast once in a blue moon without hay or grass to nibble in between. Watching ponies graze in a field, you wonder if they ever stop eating, but in fact they do for about three-quarters of the time. They rest more, and eat less, at night. This is why many owners in the spring and summer danger time, when the grass is very rich, choose to keep their animals in by day and turn them out in the evening; as well as being the coolest time and when they have most peace from flies and midges.

I do not propose to give rigid feeding tables here, as my smallest pony needs much more food than the biggest one to keep his old bones well covered. A pony's age, size, temperament, type and how hard he works, all influence what must be your decision regarding how much he's given to eat. Hence the old saying 'The eye of the Master makes the

Horse Fat'. You should get to know his eating habits and what he can look like in peak condition and aim to balance the two.

As a general rule, a pony shouldn't be given more than he can cope with at any one time. If he leaves part of his meal you're overfacing him, i.e. giving him too much. It may be that he needs that amount as part of his overall daily intake, but he will be more inclined to eat an extra, smaller amount, later. Packing masses of food into him for energy is unwise before he has to work hard. Ponies can't be exercised on full stomachs.

On a show day, for instance, we feed some 'early morning tea' in the form of hay, and a light breakfast bucket before we set off. We take hay nets for when they have finished their classes and to keep them content on the journey home, where they will have an extra nourishing bucket with the addition of molasses or glucose to counteract tiredness and put back some of the energy they've lost. Once bedded down they'll have *ad lib* hay to nibble at through the night. Inferior hay, I learned early on, is a total waste of money. Dusty, who frequently gets ideas above his station, spreads hay which he doesn't think is up to the required standard, underneath him and promptly soils it in an unmistakable gesture of contempt.

Some ponies, the native breeds especially, need very little in the way of feed to keep them healthy, just ample grazing with the addition of good hay in winter when the grass is poor and lacking in nutrients and concentrates when they are required to do especially hard work. Poor feeding soon tells in loss of condition. A pony kept in work, one which has to be show-fit and show-fat and kept in a stable for most of the time, will need not less than three feeds a day made up of a balanced carbohydrate, protein, fat and fibre combination. The easiest way of course is to feed nuts (or cubes), or a coarse mix where the manufacturers have taken the trouble to work out a balance for you, but a great many ponies fizz up on just this sort of food and you will have to find out your own combination, most probably by trial and error. Don't however chop and change feeds suddenly, anything new must be introduced gradually over several days to avoid upsetting his digestive system. Hay is the best source of fibre, stabled ponies need more of it than one turned out to grass ... and it would provide the bulk of the food weight required by your pony. It should have been made the previous year, smell sweet, be free from dust of meadow or seed type and should be crisp to touch. Unless he's too fat you can feed it *ad lib*.

When it comes to concentrates there's a choice which must start with the mention of oats, as many professional pony producers will tell you there is nothing to beat them in food value. Yet I know few children's ponies who can safely be fed oats; they become overheated and excitable when oats are added to the diet, which is why flaked maize, bruised, flaked or boiled barley, sugar beet, pony nuts, chaff and bran are more likely to form the basis of concentrated food. Sugar beet cannot be fed dry and must always be soaked for twenty-four hours in advance, but it's worth it, it seems to keep the weight on and makes other food more palatable when mixed.

Our boys have a mixture of beet, flaked barley, bran and chaff, with nuts added in relation to the work they're doing and this is fed twice a day. In the night feed they have a daily tablespoon of cod-liver oil, a good pinch of salt and a dollop of molasses – excellent for tired people, plus plenty of super hay which we're lucky to get from a farmer who takes great pride in the quality. When I go to buy, he hands me various samples to feel and to smell and there's great discussion about the various types rather like choosing cigars or good wine or a bale of dress material. When the ritual is over and the price agreed, there is never a blade wasted.

Our boys are turned out in the daytime. If they were stabled round the clock, the concentrates would be spread over three feeds and they would need more hay than ponies at grass.

Chaff is a good filler and good for the digestion because it stops them bolting their food. In the old days you had to make chaff by hand, chopping the hay, but now it's conveniently packed ready mixed with a coating of molasses in handy bags. In the summer when the show season starts, we gradually cut out sugar beet and increase proportions of pony nuts.

Frugal and fastidious owners boil barley and linseed on stoves for super condition, and I do admire them. My kitchen skills are confined to soaking the sugar beet and cutting a handful of carrots or turnips lengthways as a healthy treat for the bedtime buckets. Beet has a very limited shelf life once it's soaked, you must throw away any leftovers and make it absolutely fresh each day.

For the working pony, bran mashes given at the end of the week and before his day off, act as a mild laxative and restorative to the digestive system. They are simply prepared by soaking the bran in enough boiling water to make a soft but not sloppy consistency, and mixing in a tablespoon of Epsom salts. Cover the feed and leave it until it is sufficiently cool to eat and then feed it fresh and still warm like new-baked bread, preferably last thing at night.

A good indication that a pony is being fed the right things for his insides is in the droppings. If they are small, hard and dark it's a sure sign of constipation, ideally they should be light tan (with green overtones if he's on grass), shiny and oval shaped, breaking as they hit the ground.

4 EASY TO BOX, CATCH AND SHOE

Abbey George (age 11)

Transport

We seemed to have everything we could possibly need: a child's pony, a mum's pony, tack for both, just enough boxes and stabling for two to work them on a combined system of half-in half-out, and a map of the bridleways. What more could life offer?

'We must introduce you to Jill,' said Lorraine and Chris one day. 'Her daughter is the same age as Sam and she's pretty keen on showing ponies.' This turned out to be the understatement of all time. Jill is the archetypal pony mother, her children were crawling around in jodhpurs when others were still in Babygros. She is fiercely competitive and admits ponies are an obsession.

I blame her for a lot of the trials and tribulations in the years since we met. Her first suggestion, I have to admit, that Sam should join our local branch of the Pony Club so that we could get away from just hacking around our area, meet other pony children, receive instruction at Pony Club Rallies, and gently ease into the more competitive things, was at least a sensible one. The problem was, having joined, how did we get to all these great fun events? 'I'll give you a lift,' said Jill, which she did and it was much appreciated, but as the odd Pony Club event led to the odd local show in the summer and our diary gradually filled with horsy events as time went by, it was obvious we needed our own transport.

We sold our perfectly good car, which did thousands of miles to the gallon, and got an estate with a tow bar, the petrol tank of which I had checked twice in the first fortnight we had it – I was convinced it was leaking. The chap at the filling station and I seemed to be 'going steady' by the end of the first month.

Good second-hand trailers, i.e. ones which have solid floors, with hitching parts, wheels and brakes in good order, are not all that easy to come by. However, we had only one tiny pony to transport, so in theory it couldn't be so difficult.

'For God's sake don't get a single trailer,' said the vet, 'I've rescued more injured animals from upturned single trailers than I care to remember.' It hadn't even occurred to me that trailers capsize, but to be on the safe side we'd get a more stable, double trailer.

'No good,' said Chris, when I thought I'd found one and had joyfully taken him along for a second opinion. 'It's all wood which makes it damned heavy to pull, you'll have a job getting bits of bodywork replaced, and it's old fashioned, everyone's after fibreglass etc, you'll never sell it again.'

Five rejects later, we got one that everyone said was fine. The great day dawned and we were off. Not to Wembley or Stoneleigh but, with bales of straw standing in for a pony, to the local airfield so that I could practise without knocking down pedestrians, hitting other vehicles or just generally making a prat of myself in front of other people. Sam put down a line of spaced-out cones for me to do the bending race. I got everyone down first go.

'Try reverse!'

'Try minding your own business, it's bad enough going forward!'

As tempers frayed and night fell, tired and weary we got home, blocking the entrance to the house and half the road because I couldn't get up the drive without getting stuck.

Next day, it all seemed to get better and hope sprang eternal as I reversed all over the village losing only a couple of the school's dustbins and one of my rhododendron bushes which had never been happy anyway.

We were ready to take on an actual event. Coincidentally, Dusty was ready for his first outing at a show, and whatever else we could expect of him, he would be a pleasure to load and travel. Dolled up in his smart new going-away outfit of matching rug and boots, in he went. Oh the excitement and the joy when we got to the top of a steep hill, and then climbed out and ran back to see if he was still there, and sure enough he was, calmly munching on his hay net. Down the other side like a roller coaster at only 2 miles per hour, unable to believe we could stop at the bottom, then into the showground and, despite the boggy conditions, we made it in a nice curved arc and into line with all the ones who'd been doing it for years. I was worn out with the nervous tension, but felt I'd just conquered the world.

We got the pony ready and worked in for the 12.2 jumping. It was then we discovered why Dusty had just missed becoming tinned cat food by a whisker before we bought him.... A few feet inside the ring he stopped dead in his tracks, rigid with fear at the sight of the jumps, and no amount of encouragement would move him on. I felt heart sorry for Sam who had worked so hard up to this day, with a pony she absolutely adored. One of the pony mums suggested giving him a lead. Worth a try and this was a very 'local' local event, so outside inter-ference was actually encouraged rather than ruled out.

I led Dusty up towards the jump. It was like trying to shove a car with a flat battery, but as they gradually built up speed and I could feel his engine starting, I let go. As he went over, so did I, and most inelegantly with a great crunch flat on to my ... quarters, and a sickening pain shot down my left arm.

Getting up off the long, wet, treacherous grass I examined the damage. My left hand looked as though it was disappearing back up my sleeve. It was doing a U-turn at the wrist. Clearly we were well into injury time as a great cheer went up from the commentary box. 'Well done, Mother,' whooped the irrepressible Mrs Jacobi, 'that certainly did the trick – clear round!'

Leaving Sam to wait for the jump off, I limped off to the St Johns Ambulance folk. 'I think I may have broken this,' I said, doing my Captain Hook impersonation for the first-aid officer on duty.

'Cor, I think you may be right love,' said he, amid gales of laughter, and we drove posthaste to Oxford with blue lights flashing.

'What did you do, jump off a roof?' said the nice young doctor at the John Radcliffe Hospital, who couldn't keep a straight face either throughout the whole explanation as to how I came to break a wrist in three places push-starting a pony. As the plaster went on it crossed my mind he might not let me out again, clearly he thought I should be locked up. 'Or we might make the local rag' a nurse suggested, 'Girl Rider makes impressive debut as Mother fills the stage with cast!'

Next day, after all the build-up for the new season, Dusty was put out to grass for at least six weeks until my plaster came off, and the trailer, which I'd finally learned to tow, was up for sale. At the time it seemed that driving a horse box with a reconditioned wrist would be easier.

Trailers aren't as simple as they may at first appear. They still have to be serviced annually, just as a horse box would, for tyre checks and brakes in particular. Sitting out in all weathers, brakes can easily go rusty. A sound floor is essential, so many ponies are lost through accidents when a floor-board gives way.

Whatever you travel an animal in, it should be mucked out immediately at the end of the day. Soiled bedding must be removed, the clean stuff brushed up front and floor stains washed off before they can rot the wood. Rubber mats are a useful extra, they keep the floor clean and provide the pony with a non-slip foothold. Bolts, hinges and other movable parts must be kept well lubricated and tyre pressures need regular checking so that the trailer doesn't pull unevenly or zig-zag.

The law lays down a good deal regarding trailers, of which horsy folk are often blissfully ignorant. You must have illuminated number plates, rear lights and rear reflective triangles, brake lights and indicators. You must have brakes on ALL wheels, in good working order at all times. After you reverse, you must disengage the manual lever before entering a public highway. Otherwise your brakes won't work and you could be prosecuted. There are speed limits restricting you to 40 mph when towing, 50 mph in certain cases, but you must display a 50 mph plate on the rear of the trailer.

Obtain a copy of 'Speed Limits for Caravans and Other Trailers' from HM Bookshops, and further information from the Department of Transport, at 2 Marsham Street, London SW1 (Tel. 01-212 3434), or put yourself in the hands of a specialist at a trailer centre to do the necessary checks and give you advice before you even venture forth.

Most important, it is against the law to travel in a trailer with ponies or horses, whatever small children say about it being perfectly all right. For most women drivers especially, a horse box is more manageable, but you do have to pay extra for road tax and insurance. Our box, which is big enough for two ponies, can be driven on an ordinary driver's licence. It's designed so that the ponies travel sideways, and they can therefore be turned easily to walk out head first down a ramp. The usual travelling position in trailers and horse boxes comes in for a lot of criticism and yet horse-carrying vehicles still continue to be designed so that the animals are facing forward with their heads tied at the front, which means that their body weight is pushed back on to the hindquarters as they are driven. Animal behaviour experts say it's a fact that ponies prefer to face the back, and are happy watching the countryside that has gone past. Standing sideways is the next best thing for their peace of mind and comfort. The good news about both trailers and horse boxes is that because they are always in great demand, they tend to hold their price very well – unless you are extremely negligent about maintenance.

Easy to box

This is generally tacked alongside other fibs such as 'sadly outgrown' and 'always in the ribbons'. A pony who causes loading problems can put a damper on a day's outing to a show, Pony Club event or whatever. People will tell you that the more often a bad boxer is loaded, the less resistant he will become. Ignore them, the situation can get worse. Our Gay doesn't load at all. If she qualified for Peterborough locally tomorrow we would have to walk the 100 miles to get there.

When we got her she wasn't exactly enthusiastic about loading, but by letting her stand on the ramp for a bit whilst she pondered the situation and we hummed a few bars of her favourite tune and pretended we had all day to hang around, all of a sudden up she would toddle and off we would go.

When my arm was in plaster and I couldn't drive, a seemingly knowledgeable horse person agreed to transport Gay back from a farm where she'd been turned out. Apparently, as she stopped on the ramp to listen to the birds and say goodbye to all her fans, an ill-advised attempt was made to stuff a yard broom head first up her backside to hurry her on. When she was home again, she would flee to the back of her stable every time a broom appeared. Out on rides she would shoot forward at every sudden movement from behind – a newspaper boy coming behind on a bicycle nearly landed me on to a pile of sausages in a butcher's van, and innocent joggers bringing up the rear on country roads stopped in their tracks as a clearly frightened pony hurtled into the air in front of them. We discovered the reason for all this when we loaned her to a teenage friend to take to Pony Club camp. At the sight of the trailer she shot into reverse. We could have berthed the *Queen Mary* in the time it took to get her in a straight line, all the while becoming more distressed and more dangerous.

We save our friend Chris for such a crisis. Chris never gets rattled with horses, they love him at first sight. At the time he'd sold me Gay a few years earlier he deposited her with the parting words. 'And if you ever find she's too much for you, let me have her back and I'll get you a rocking horse.' He'd soon sort her out.

All the normal Gay persuasions having failed for him too, he tried the broomstick method. The pony went wild, rearing up and spinning round converting the broom to a bundle of firewood and removing Chris's hat and almost his front teeth, all in one movement as Lorraine flew elegantly through the air – like Peter Pan on the end of a lunge line. Ages later with the pony and all around exhausted, we got her in but it was not a pleasant experience and I was intensely relieved to hear she'd unloaded calmly and safely at journey's end.

Poor Abbey fell off a ramp at the first attempt to load her when she was very young and it stuck with her a long time. Nowadays she walks calmly in beside her beloved Dusty and would travel forever with him by her side, but when there were problems originally, again it was a quiet and confident man who persuaded her gently with the aid of a lunge line attached to the side of the box.

Beware the difficult loader. A pony like Dusty who trots up a ramp in his eagerness to go places and who instills total confidence in any nervous animal who shares the vehicle with him, is heaven. When there isn't another pony, a goat is a popular choice as a travelling companion – several racehorse trainers take a nanny goat along with nervous youngsters. I know this solved the loading problem for one pony mother, when her pony went up the ramp and stood quietly the many miles to a county show with its little nanny alongside. Unfortunately, when the pony was unpacked the other end it was minus its row of beautiful plaits. A goat's passion for junk food knows no limits it seems.

Time and patience are the operative words when loading. Both tend to be a bit scarce when you have a deadline to meet. Here are some of the methods which I know have worked for other people.

Youngsters are often encouraged to load by feeding them in the trailer every day for a week and this can do the trick with difficult loaders. Gradually the ramp can be put up and the pony driven a short distance, then perhaps let out for a short hack before returning home.

A strange, cramped and dimly lit interior can put ponies off. Removing the partition and covering the floor with straw, making sure the ramp is solid and not slippery, and, if there's a front ramp, letting that down to create an impression of light and space, all helps.

Leading him in in a bridle gives more control than a headcollar. Many a persistently bad loader succumbs to the cross-lunge-line method (see Fig. 1). You need two lines attached to the ramp fastenings. As he is led up the ramp, two helpers cross over behind the pony, gently easing him in. Perfect where you have a small battalion on call, no good at all for a lone rider or a single parent with a small child.

Try a single-lunge-rein method (see Fig. 2). Place the trailer against

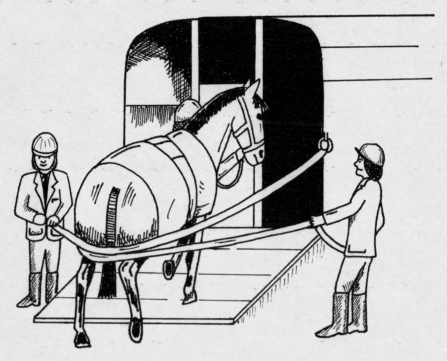

a wall, so that he can't hop off the ramp at the right side. Dress the pony in a lungeing cavesson and attach a lunge line, the loose end of which you tether through a ring inside the trailer at the front. In theory, when you lean on the line, this pulley method encourages the pony to go forward as you take steps back to maintain the tension. Barbara Woodhouse and others swear by this method. The muscles in his neck should give up the strain before the lunge line does.

It must be said that not all ponies are Gay, and many have been successfully persuaded to load with the aid of a short sharp prod in the rear from the bristles of a broom or a flick with a piece of brushwood. Don't in any circumstances be tempted to hit him. Getting annoyed, shouting and struggling only increases the tension in pony and handler. It takes two to pull. If you don't, he can't. Equally, handlers who are nervous and hesitant about loading convey their uncertainty to the pony. A positive but calm approach wins. This is why professional transporters who regard loading as all in a day's work, getting on with it with the minimum of fuss, rarely have problems. If you can afford to, letting an expert handler take the pony a few times could give him

Fig. 2

confidence and cure him of bad habits. Incidentally, it's against the law for a private owner to give a lift to other people's ponies, although it happens all the time. There could be problems with insurance claims in the event of an accident, so it's worth checking to find out if you're covered, wherever or with whoever your pony travels.

Whatever the loading methods, a great fuss should be made and a reward offered immediately, once he's in. Because I'm all for a quiet life, I go even further with our boys and they have their breakfast buckets once they're loaded and whilst we're putting the finishing touches to the packing. Professionals load everything they need first and the pony last, but this meals-on-wheels method works for *us* and starts us all off in a good frame of mind. It doesn't seem to upset their digestion and it sure helps mine.

Easy to catch

To catch a pony all you do is walk into its field, carrying a headcollar and leadrope over your shoulder, and call his name. As he comes obediently towards you the instant he's asked, with your left hand reward him with a titbit for being a good boy and with your right, slip the rope under and over his lower neck. Reaching surely but unhurriedly for the spare end, bring it round, making a loose noose and

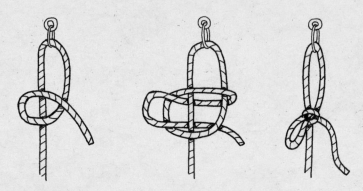

Using a quick-release knot is the safe way to tie up a pony.

you've got him. He'll wait until you fix the headcollar on and away you go. That's the theory at any rate – and some little gems will co-operate, indeed will almost knock you down in an effort to be caught, especially if they associate it with something nice, like resting their legs in a deep warm bed, or their noses in a bucketful of food.

Others will give you the run around and drive you to distraction as you chase after them. Don't. They enjoy this game of catch which you are playing their way if you run after them and get agitated. Remember the old Chinese proverb about everything coming to he or she who waits, and, next time you're waiting, try a few crafty tricks to fox him. For a start, you could just sit on the gate and produce from your pocket a rustly brown paper bag of mints which you slowly and deliberately feed to yourself. This will quite often do the trick and bring a greedy pony up to you, as indeed will a bucket of nuts. However, fat, full and wily ponies will not be tempted with food which is why many experienced handlers prefer to use water as the bait. The pony is deprived of his constant water supply, and has water taken to him twice a day in a bucket. Once he has adapted to the new routine, he has to allow himself to be caught first and then he can drink, with the handler positioned to the side so that she isn't rewarded with a smash in the teeth when he suddenly raises his head. After he's been caught, he's allowed to drink his fill from however many buckets it requires whilst his handler goes through the 'what a good boy' routine with rapturous applause, and turns him loose. If it takes more than a quarter of an hour to catch him by this method, you have to take the water away and bring it back a few hours later, and then at shorter intervals after that.

If a whole day goes by and he's absolutely parched when he's caught, you would of course have to restrict his intake to half a bucket, followed by another half, an hour later, gradually building up to an unlimited supply. This method doesn't appeal to me because of the time you need to make available to do it properly and because I wouldn't want to suggest to children that they could interfere with a pony's water supply. But I am told it is very effective and anything is worth a try to end the misery of keeping a pony you can admire only at a distance in a field.

I prefer Wilma's method which always does the trick for her. She casually walks into a field and starts looking for something she's lost in the grass, ignoring the pony completely. She then walks gradually in an ever-decreasing circle until she's quite close, but still not acknowledging him at all. At this point when he has accepted that she has come on entirely different business, I have seen ponies so overcome with curiosity, that they walk over to her and join in the search. She then just puts a hand out and rests it on his headcollar, and still looking for the lost object they stroll towards the exit. It sounds unlikely but it never fails. There is always a hazard to ponies turned out in headcollars that they may catch them on protruding objects, but they do make catching a lot easier, and to leave about 6″ of rope or twine attached to the headcollar under his chin will simplify matters even more.

Easy to shoe

'What will you do', I asked Albert Berry, our blacksmith, on the day of Abbey's first shoes, 'if she won't let you get them on?' 'I'll say tattie bye, and hope you get on with her as best you can,' said Albert kindly as Abbey backed and snorted and looked like giving him a hard time.

For her first shoeing, Chris held her head and talked to her as Albert worked. Ever after that she just stood and had forty winks as the others do, with just a rope draped over their necks while their feet are trimmed and shod.

A good blacksmith will tell you that he rarely has trouble, and again I am sure it is all to do with trust. Some owners half expect problems and if you're a nervous type it's best to keep out of the way, unless your presence is requested.

In days gone by blacksmiths not only dealt with feet but also with the teeth – rasping, pulling out wolf teeth etc. Colin Hill, who is our horse dentist, learned the trade from a blacksmith. When he first came to do the boys I turned ashen at the sight of his tools; the contents of his bag looked like the spoils from a raid on a torture chamber. His only request was for a bucket of clean water to wash them in as he worked. I waited for the ponies to exit backwards, as in Tom and Jerry cartoons, clean through the walls of their stables, but once they got used to the idea they just stood quietly as teeth were rasped or pulled. 'You shouldn't expect animals to feel the emotional way humans do over illness or injury,' a vet in the Middle East once reprimanded me as my cat had a leg removed, 'Don't you dare convey your anxiety to her.' So, as the teeth came out, I tried to avert my gaze from the bloody bits by examining Colin for scars, but no, he hadn't got any. 'It's how you approach them,' he assured me, 'If you're firm and confident, they won't mind.'

(A pony with a foreleg held up can't go very far – a useful trick employed by handlers when they need a pony to stand still for clipping, applying a dressing or whatever.)

Whether a pony is ridden or not, his feet must have attention at least every six weeks, and most will have become used to a blacksmith trimming their feet at regular intervals from an early age. The mistake which many less knowledgeable owners make is to judge from the condition of the shoes whether or not it's time for the farrier, regardless of whether the pony has developed corns where the shoe presses into the foot, or if the horn is cracking, or if the clenches holding the shoe are protruding from their holes. Rather than wait until the shoes wear paper thin or start clanking and falling by the roadside, ask your farrier to do a 'remove' every four to six weeks. This means he will, if the shoes are in good condition, trim the feet and re-bed the shoes.

It is inadvisable to shoe before the age of three; it gives the feet a chance to harden, and once you start you have to keep up the shoeing – the feet become dependent on them. I like to turn our ponies out for their annual holiday without their shoes, but some owners prefer to keep shoes on to protect the feet. If ponies have been unshod for a while some blacksmiths may tell you to do a little work on hard surfaces to knock the edges off and toughen them up again. It is best to take the advice of your own farrier here.

If a pony is impossible to shoe it's almost certain to be because of a bad experience earlier in his life, and you may never put him right. In which case take Albert's advice, and tell the vendor 'tattie bye'.

Anyone trying to sell a pony with any of the aforementioned problems might just be tempted to cover them up before a prospective purchaser arrives. For your own protection, therefore, arrange to see a pony brought in from a field if possible to see how easily he can be caught. Ask if your child can put the saddle and bridle on, to get an idea of how well mannered he is to handle in a stable. Will he allow your child to pick his feet up? Can she mount him without assistance or does he fidget and move forward unless held? If he's with stable companions when you call, does he leave them quite happily to go and do his bit for you in a field or wherever? Ponies which nap can frustrate and exhaust children. To test him in traffic is tricky, but it's essential to know that he's safe on the roads before you buy. You need a test drive with an adult accompanying him past tractors, buses, motor bikes, big lorries and anything else you can arrange. (See Road Tests page 94.)

We will assume that when your child rides him the pony doesn't flatten his ears back against his head or display other signs of unwillingness, but to be fair, you should allow for a period of slight adjustment on both sides. After the ride, ask to see him taken to a trailer or horse box – your own preferably – to see how he's going to load for you. You can understand why it's desirable to have a pony at your home for a trial period; it's good Public Relations on the part of the vendor and inspires confidence in the prospective buyer, as does having an experienced person at your elbow for a second opinion when you go to look him over, but it is by no means a guarantee. If there is the slightest doubt, sleep on it. Remind yourself that there are more decent ponies looking for homes than there are decent people with the means to buy them. As we discovered with Dusty, you can't know everything about a pony's past during a loan period, and you can't hope to put right everything you feel has been caused by bad upbringing. He's given to bouts of temperament like all talented artistes at times. He's inclined to think he is Noël Coward, but we allow him a bit of swank, it's part of his character.

A friend of mine bought a working hunter pony in the summer when it was thoroughly ladylike at home, and did everything that was asked of it at shows. When winter came and a pair of clippers was produced to tackle its thicker coat, it turned into a werewolf; it could be clipped only under an anaesthetic it seemed. Time and patience have been rewarded in both cases but, alas, where small ponies change hands at such speed, due to a child's growth and through no fault of their own, time is often the one thing you don't have.

5 ON MATTERS OF HEALTH

Samantha Baguley (age 13)

'Have you cuddled your vet today?' asks the sticker on the back of the car driven by one of our most revered Pony Club instructresses.

You may hesitate to call a vet because you feel the problem with a pony is too minor to waste his time, or you may feel reluctant to spend money where you feel you could treat the pony yourself. Both attitudes are potentially dangerous. I have learned the hard way, and now, however minor a problem, I always ring the vet to describe any changes in a pony's condition which are giving cause for concern, and to ask him what can be done. Obviously, unless in an emergency, he will not be too thrilled to hear from you at two o'clock on a Sunday morning in his bare feet and nightshirt. And at times he will decide a visit is unnecessary and advise a home treatment, but a good vet will always tell you to call him back if the condition doesn't improve in a specified time. Vets can often become a little tetchy about owners who call them out for one thing and then ask for a 'job lot' of examinations on other animals all round the yard, so it's wise to specify all that the visit may entail and not spring it on him when he has allocated only ten minutes for a vaccination. Early diagnosis and treatment can prevent a more serious and costly complication later, so pay gladly whatever a vet asks – his years of training and your peace of mind are worth every penny. If you think *his* fees are high, perhaps you haven't had to call out a plumber or a TV repair man recently.

First-aid kit

It makes no sense to have all the things you need to clean up a wound sitting at home, when your pony cuts his leg at a show, so it's a good policy to duplicate first-aid stuff and always travel with a spare box wherever you go with your pony in case of an emergency.

A basic kit would include:
Disinfectant
Cotton wool
Large jar of Vaseline or zinc and castor oil
Antiseptic dusting powder in a puffer bottle or aerosol can of antibiotic.
Non-stick dressings
A roll of gamgee tissue
Sharp (but round edged) scissors
Crepe bandages

Kaolin poultice dressings (in sealed foil envelopes)
Thermometer
You will soon get to know when a pony is looking his best in terms of health and fitness and when something is wrong with him. Prevention, as we all know, is better than cure, and, since many pony illnesses are avoidable, a sensible regime of good food, proper exercise, good grooming, plus regular checks of his feet and teeth, and diligent worming should prevent the more common complaints which arise through neglect or ignorance.

A child who is old enough to have a pony, is old enough at least to help out with the animal's general care. If there are constant complaints about the amount of work involved or displays of reluctance to ride when weather conditions deteriorate, if friends who don't ride constantly call round, or it's the night for the Brownies, playing the recorder, singing in the choir, or other pastimes which need priority, then it might be time to think of a new home for the pony where he will be better appreciated. It is heartbreaking to think of how many genuine pony-loving children there are who will never get as far as a riding lesson, and how many spoiled brats actually have to be bribed to take an interest in the unfortunate animal who has been acquired like a BMX bike or a portable television.

Children, mercifully, do not have the means to go out and buy ponies and provide care and accommodation for them without adults assuming the ultimate responsibility. The worrying thing is that so many ponies are left in the sole care of children who are too young or inexperienced to cope. It is amazing how many adults will accept a child's word about their riding ability and knowledge of ponies. Perhaps they assume it's just like keeping a goldfish or a rabbit, except that there is quite a bit of money involved, so it's a status symbol to boot.

Pony Club membership is a good investment for novice parents. Pony Club mothers will fall over backwards to lend a hand where the welfare of a pony is at stake (the child may be dying of the palsy, but so long as the pony is okay!). Pony Club instructresses are quick to spot if a pony is too spirited for a beginner to handle, if a pony is off colour, if he needs a change of diet, new shoes, a different type of bit or whatever. Most Pony Club branches are crying out for helpers at rallies, camp, or competitive events, so parents and children can become involved in the various activities. Pony Club exams quickly sort out how much a child is concerned with stable management and health issues, and never mind that she looks cute in the saddle.

Obtaining a few reference books (see Recommended Reading on page 141) and the seeking out of a first-class vet, blacksmith and food supplier, are parents' duties, even before the purchase of a pony is complete.

Vaccination

Pony Club ponies must be vaccinated before they can attend camp, and many shows require proof of inoculation against equine influenza and tetanus. A pony receives a booster vaccination annually. When being vaccinated for the first time, a pony has two initial doses separated by a four- to six-week interval. Your vet will tell you how long to rest the pony after vaccination, but it can be anything from twenty-four hours to one week. Most ponies appear not to be affected by vaccination, but others feel very sorry for themselves, hanging their heads, feeling very tender around the neck where the jab was given, losing interest in food and even running a temperature.

Health indicators and common ailments

The normal *temperature* is 38°C (100-101.5°F) and the pulse rate 35 to 45 beats per minute. You shouldn't be able to hear a pony breathing, but a slight movement in his flanks indicates his breath going in and out at anything between 8 and 12 times per minute. A pony who goes off his food and seems to lie down more often than usual, especially when there are visitors around, could well be feeling ill. At such times you may also notice a dull look to his eyes and a 'staring' coat – the hairs on an unhealthy coat stand up instead of lying close to the body.

Teeth. The reason you can tell a pony's age from his teeth for most of his life is because they keep growing, but they are prevented from becoming tusks by the fact that they are worn away by rubbing against the ones in the opposite jaw as the pony eats. Because we do not stick braces on them as we do with children, where they grow unevenly due to irregularities within the jaw, they do need dental help, so at least a once-yearly check of a pony's teeth is essential. If overgrown or 'sharp' teeth are left unattended, the pony won't be able to eat properly and his digestion and general condition will suffer. Amazing improvements can be seen when a pony with neglected teeth is put right, which is why our Gay has her teeth done every year at the onset of winter and not in the earlier part of the year when the lush grass comes in. Her dentist told her lately that she had quite a nice mouth for a teenager and that considering her waistline, if her teeth were any better she would explode!

Feet. Nearly all lameness stems from the foot. Common problems being pus in the foot, bruising or of course laminitis. Some nursing mothers put sore pony feet in buckets of warm salt water solutions, others tie bags of wet bran to which Epsom salts have been added, and both these methods seem to reduce inflammation and relieve pain.

Regular attention from a blacksmith, whether or not the pony is wearing shoes, is one of the main priorities of pony care. Daily cleaning to keep them free from dirt and mud, and fresh dry bedding to stand on, will help avoid diseases like thrush, detected by a foul smell emitting from the soft horn of the frog – the V-shaped cushion surrounded by the sole. Neglect is the most common cause, so a radical change in the way the pony is cared for is required.

If a pony's shoes are allowed to pinch, he can develop corns, but leaving him unshod and left out in mud and rain can bring problems of a different nature.

Cracked heels don't, strictly speaking, mean cracked heels, since the affected area is higher up, in the hollow behind the pastern. When the skin becomes wet and muddy, and consequently chapped, it can also become red and tender. Vaseline works well as a barrier cream, but once the condition is advanced – with the ensuing itchiness followed by pain for the pony – smear the area with Stockholm Tar and take your vet's advice. Once the skin cracks open, germs can enter and the condition can take weeks to heal, so antibiotics may then be necessary. The way to prevent this is to make sure the affected areas are dried off thoroughly when he comes in from exercise, and by using a barrier cream to protect him against both very dry and wet muddy conditions when he's living out.

Mud fever is an infection which occurs in the lower limbs in particular affecting the skin behind the pasterns. It is caused by chapping, again as a result of cold and wet conditions, and if left unchecked the pony's legs will soon swell up and become hot. This is one very good reason why you have to check a pony's movement when he's turned out in a field for the winter, and not just wave to him from the school bus or on the way to Tesco's.

An antiseptic ointment may give some relief if the pony is brought in and scabs are removed to give the skin underneath a chance to heal, but in more severe cases, where he is apparently lame, you will need to call the vet. This is another condition that can be aggravated by dirty bedding.

Laminitis. The scourge of small, fat, greedy ponies. Although in theory you should provide two acres of grazing land per pony, this should allow for fencing off half the grazing area in summer, particularly when too much lush early summer grass can send half of them foundering. So restriction of food intake is the first step where a pony is prone to laminitis; once he's had an attack his chances of contracting it again are greatly increased.

However it isn't just grass which causes this complaint. Stabled ponies, although less commonly, can catch it too, especially when their diet is too rich and their intestines become overloaded with carbohydrate, or if they have had too much work and are suffering from exhaustion, or have had a reaction to drugs prescribed for a different infection. This is how Gay managed to get it in mid-January, when she had had medicine for a mild attack of bronchitis. The majority of cases can be prevented by not overfeeding – even too many pony nuts can cause a pony to become laminitic. Providing him with extra hay if he's hungry is always the wiser course, with regular exercise sessions to stop him going to fat. A working pony is a happier pony than one who is left to stand and nibble all day long to relieve boredom. One hour at grass each day is enough in summer if the pony is the laminitic type, but should you ever find him leaning back as if trying to take the weight off his forelegs, or if you feel heat in the front part of the foot as you examine it, then the condition is already advanced enough for you to take drastic steps to treat it. Get him away from grass and any other food that's around, call a vet, and in the meantime hose down his feet with cold water to ease some of the pain. If you catch it early enough, allow the alimentary tract to empty by giving his stomach a complete rest, then gradually introduce a restricted diet of small amounts of hay and bran mashes to which you can add Epsom salts and a handful of roots to brighten them up, and take him walkies as often as you can.

A pony at rest may raise one hindleg and stand with it bent as he relaxes, but if he does it with a foreleg it's an indication that there's something wrong with it.

Lameness. The causes may be many – pus in the foot, a tendon injury etc., all of which will need immediate professional diagnosis.

When a pony is trotted out on a loose rein he lifts his head as a lame foreleg strikes the ground, and nods it down again as a sound one follows. Lameness in the hindlegs is detected by watching the movement in his quarters. They lift as the lame leg strikes the ground, lower as the sound leg makes contact. Perhaps it helps to remember which way round it is if you think of him pulling up, away from the pain, and relaxing down, when it's easier.

Coughing. Not a welcome sound. Persistent coughing, untreated, can lead to permanent lung damage; coughing can be the first symptom of broken wind. On the other hand, some ponies cough because of dusty straw or hay. Spraying the bed to damp it down after you lay new bedding, or changing to wood shavings, or dunking the pony's hay net through a water trough before you hang it up, and administering a teaspoon of honey to ease the tickle will all help. Don't take any chances with coughs – they can quickly spread infectious diseases. On no account take a coughing pony near others, and make sure his buckets are kept scrupulously clean and not used by his stablemates.

Cuts and grazes do not heal as quickly on moving parts of the body. A vet should be called where a wound is of any depth i.e., right through the skin, and may require stitching. Nearly all wounds are contaminated with dirt when they occur, so the first action must be to clean and disinfect the injury to stop infection.

Colic. Abdominal pain. The pony may paw at the ground with his forelegs, keep bending his head round to look at his stomach where the pain is located, or lift his hindlegs as if to kick his stomach in an effort to remove the cause of his discomfort.

Dusty had a spate of colicky bouts early on when he had had a few intestinal problems due to malnourishment. It was difficult to put weight on him, because extra food, unless in tiny amounts, gave him a stomach ache. He always wanted to lie down, which bothered me as I had heard awful tales of twisted guts, which can of course happen in chronic cases. Our vet said that unless he was rolling or thrashing about in a manner likely to cause him injury, it was all right for him to rest quietly in between regular walking sessions. So every half-hour, dressed in warm rugs, he was led through the village for a ten-minute walk (sitting down on his haunches like a dog on one occasion and causing a traffic jam outside the post office by refusing to budge) then he would come back and have another lie down. With his head resting pathetically on Sam's lap, he had a story read to him and his ears gently pulled from root to tip, which always seems to comfort him. Suddenly he'd stop gazing mournfully at his sore stomach, and milking every drop of sympathy he could get from the situation, and up he'd get as perky as ever. We naturally took away all food until the attack was well past, then put him on a restricted convalescent's menu in accordance with the vet's advice. It's also important to keep a nice thick bed round a pony with colic so that he can't hurt himself when he rolls.

Choking. Get immediate veterinary help as a pony can't be sick, and a muscle relaxant in the form of an injection in the neck will possibly be necessary to remove the obstruction. Again, when it clears, the transformation can be magical.

Tail rubbing. Another of our specialities, unfortunately. Abbey decided to take this up on one occasion and turned her beautiful cream tail into a lavatory brush overnight. In her case, itchiness caused by too rich a diet was diagnosed when she'd been working very hard, was ultra-fit and we'd been stepping up her rations to compensate for energy loss. In fact Abbey has a high energy level and being a good doer, she does very well on *ad lib* amounts of good hay and small non-heated mixes to which we add a teaspoon of salt. We also have to cut sugar beet out of her food completely when the grazing is good.

Applications of Benzole Benzoate (sold by chemists) give instant relief and help the healing process, but if feeding isn't the problem, consider the other common causes of ichy tails and manes.

Worms. The need for regular worming can't be overstressed (see page 32). Change the brand of medicine so that a resistance doesn't build up. Pin worms can cause irritation round the dock area.

Lice or *nits* (a word children adore and can identify with more easily). Apply delouse powder liberally according to the instructions, weekly or fortnightly for at least a month and whenever necessary thereafter, because ponies out at grass can quite easily become infected. Keep grooming tools spotless. Dirty brushes, and ponies rubbing against each other, quickly spread lice. If an itchy condition persists for any length of time, a more troublesome condition such as Mange, now comparatively rare in this country, could be the cause, but much more common is sweet itch.

Sweet Itch is an allergic dermatitis, which will need the vet's attention. It is caused by midge bites where there is a reaction to the midges' saliva. Midges are most active in the early hours of daylight and the last two or three before nightfall, so stabling the pony at such times to prevent him being bitten is obviously advisable. Your vet will suggest suitable medication such as a concentrate called Stomoxin. Sweet Itch Ease, which contains Benzole Benzoate, also helps to promote regrowth of the hair, and has the addition of lanolins and oils in the formula.

Azoturia is often referred to as 'Monday Morning Disease', basically because it is caused by bad stable management, in particular not cutting down on food when he's out of work and resting. It is a painful stiffness affecting the muscles of the loins, causing the pony to stagger when he's taken out on exercise. All exercise must then be discontinued and the pony transported home and kept warm until a vet arrives. Sometimes this condition particularly affects geldings, especially when they have to undergo long journeys in a trailer or horse box where they are unlikely to stale. Usually though, it's through not following the basic rule of feeding only in accordance with the amount of work they do. Azoturia is a serious complaint in that it physically damages the muscles.

Girth galls are caused by lack of attention when cleaning tack, or when dressing the pony. A dirty girth or one too tight or too loose can result in friction on a particular spot, rubbing away the hair and infecting the skin. Dab the sore with surgical spirit and rest the pony until it is completely healed. Once the skin is thoroughly hardened, Vaseline or zinc and castor oil ointments promote the regrowth of hair. The same treatment applies to saddle sores caused by ill-fitting saddles or rollers, in which case they must obviously be replaced without delay and not used again when the sore has been treated.

Footnote It may sound silly or sentimental to treat sick animals in the way we treat sick humans; visiting them and saying nice things to cheer them up, but ponies love gentle touching and gentle voices – even the soothing noise of a radio in the background to keep them company is said to have a beneficial effect on them. I have often marvelled at the bedside manner of our vet, who talks to the ponies as though they are the most cherished people in his life. Now for all I know he may go home and talk to his wife as though she's a horse, but it seems to me that if more people tried the same technique on animals and humans alike it wouldn't be a bad thing at all.

Feeling a little horse!

Lorraine Drewe, 1987

6 GROOMING

Caroline Lippmann (age 11)

How much grooming a pony needs will depend on whether he's living out, or whether he's stabled and rugged for at least part of the time in the winter, plus how much exercise he gets and what sort of diet he's on.

A basic list of grooming tools would include:

A rubber mitt or rubber curry comb (a priceless gadget for hairy ponies) for removing mud or loose hair from a thick coat as he sheds his winter woollies

Dandy brush

Metal curry comb

2 body brushes (or 1 body brush, 1 water brush).

Rubber

Sponges

Hoof pick

Hoof oil and applicator brush

Mane and tail combs

Sweat scraper, which also removes gallons of water at bath time.

Keep all your grooming tools in the kind of box with a handle which most supermarkets, hardware and D.I.Y. shops stock for shoe-cleaning stuff and other bits and pieces. The best time to groom a pony properly is after his daily exercise when he's warm and relaxed, but before going out he should be 'quartered', i.e. given a lick and a promise with a dandy brush to remove any bits of bedding or dung stains from his coat. Apart from improving the skin tone and helping to build up muscle, brushing the coat aids the sweating process that regulates the body temperature and gets rid of toxins through the pores.

Whether to allow mud to dry after exercise before brushing it off, or whether to hose down the muddy areas, seems to be a subject for debate. The main argument in favour of simply brushing without the use of water, is that it prevents the heels from becoming cracked and the legs chapped. On the other hand, if after a quick wash, the heels and legs are dried with clean straw and old towels and kept warm with stable bandages, the risk should be minimal. If showing, you would certainly keep his legs clean by washing and applying gamgee tissues under leg bandages so as not to mark them.

Feet should be picked out at least twice a day; that's the first and most vital part of grooming. Facing towards the tail and running your hand down the back of each leg, take hold of the foot around the front of the pastern, lift upwards and say 'Up' as you do so, he'll soon get the message. Pick out the feet thoroughly, working from heel to toe and make sure that all the dirty bits are gathered up in a container as you go along, otherwise he'll merely tread in them again as soon as you finish. A good tip here is to bore a hole through the rim of an old plastic washing-up bowl and tie a hoof pick to it with a length of twine. This way you not only have an easily portable rubbish collector but you won't lose the hoof pick in the bedding.

Usually you would remove the muddy bits from legs and belly with a dandy brush, but if he's inclined to be ticklish, substitute a body brush for in between his legs and for dusting down his family jewels, holding his tail up high to prevent him kicking. Brush out the tail (body brush again) separating the hairs with your fingers to remove any tangles.

One of the most neglected areas is underneath the mane, so lift each section and get right down to the roots where grease and scurf can gather. Take the body brush in one hand and the curry comb in the other. Standing at the nearside of the pony, start working towards his quarters from the top of the neck, using a smooth, circular, sweeping movement in the direction of the hair. It's far less tiring to stand back and use all your weight than to push from the elbow. After every few strokes, draw the brush across the teeth of the curry comb which you should then tap on the ground to dislodge the dust. Move to the offside and repeat the process.

Good grooms include a daily strapping session in their grooming routine, to build muscle and improve the circulation. This does wonders for ponies with dull coats, scraggy necks etc., but you must not touch the head, belly or loins. You will need a wisp made from dampened straw, or a strip of towelling wrapped round your hand, or a cactus cloth or leather pad designed for the job and available from saddlers.

Working on the neck, shoulders and hindquarters, thump firmly and rhythmically – the object of the exercise being to build up muscle tone as he involuntarily tenses himself in anticipation of each slap. It isn't as harsh as it sounds, and you'll both feel very good after this bit. Next throw a stable rug over him to keep him warm as you work on the head. Some ponies hate having brushes around their eyes, and that being the case you can make life easier by using a sponge or an old scrap of towelling rolled up and used as a face flannel. (Keep sponges in different colours using one to wipe over the eyes and muzzle, the other for the dock end.) Always make sure you dry his ears thoroughly. Although my preference in general terms is for geldings, give me the girls every time when it comes to grooming. Ever since Dusty withdrew his sheath when the telephone sounded loud and shrill in the yard one day, causing the sponge I was using to disappear into his person like an elephant stealing a bun, I've been slightly apprehensive about washing him around these parts. The fact that he's extremely ticklish, mincing from one leg to the other, doesn't help. Yet I see it as the cross we have to bear and performing this ceremony every week is still more agreeable for both of us than letting dirt accumulate over a longer period. Old grey gentlemen I'm told are particularly susceptible to problems but any gelding relies on his owner for regular cleaning. After washing and drying, a light smear of liquid paraffin acts as a moisturiser and makes it easier to remove the dirt the next time.

Grooming is completed by a final polish with a rubber, and the laying of the mane and tail with a damp water brush. At this point a tail bandage can be applied for a couple of hours to train the hair.

Oil the feet every day to prevent cracks and promote growth of the horn. Again, it's a controversial issue here whether to oil the underside of the feet. In fact sealing off moisture and air, which are essential for a healthy foot, could also lead to problems. Regular picking out of the feet to keep them clean seems to be the safest course, but the occasional underseal won't do any harm, and we always oil top and bottom for shows. Hooves which become clogged with dirty bedding can easily result in a nasty smelling case of thrush, and even good clean wood shavings or sawdust can accumulate around the frog and dry off the soles very quickly. A pony's natural habitat, a good grassy well-drained pasture, is the ideal one.

Pulling and plaiting a mane

PULLING

Removal will be easier and will cause less discomfort to the pony if the job is spread out over quite a few sessions. It's a shame to see good-natured animals put upon because they 'don't mind' anything. In fact they do, they're just more considerate than their owners who take advantage of them by yanking out fistfuls of hair the day before a show.

The over-all appearance will look tidier, and the mane will lie flatter against the neck if the hair is pulled from underneath. Where it has grown really thick and unruly, it will help to separate the bottom layer from the top by using hair grips or rubber bands to anchor the top layer over on the nearside as you work away on the undergrowth. Briskly pull just a few of the longer hairs at a time, wrapping them round a comb or over your index finger (which you have wisely protected with sticking plaster or a rubber glove). Although you may mean to be gentle, pulling slowly bothers the pony more than a short, sharp session followed by much praise for being patient and good.

Gradually proceed carefully with the top layer, back-combing small strands to produce an over-all depth of four or five inches. If you pull the top line, a regrowth will show. The line of the mane must be even and of uniform length from head to withers, otherwise the plaits will be of different sizes.

There is a golden rule about never touching a mane with scissors (except to remove about an inch from behind the ears where the bridle headpiece rests), and yet you'll often see people at shows snip off the odd hair that is spoiling the beautifully sculptured look of a plaited neck. I have to confess, because Gay hates being pulled and because nature gave her a mane like an overgrown hedge, I keep it in trim with the kind of thinning razor which hairdressers use and which you can buy at the larger chemists. It doesn't leave a straight edge and it really works very well to keep her thick New Forest tresses from flowing straight down to her knees.

At the end of a pulling session, lay the mane with a damp water brush. A thick mane tends to flip over and fall on the wrong side. Train it by damping it down, dividing it in sections and leaving it in bunches or plaits in rubber bands from time to time until it lies naturally on the offside again. Then use tubular, stockinette duster cloth pulled over the head as a hair net. Sparse growth on manes and tails, and any hair which breaks easily, can be improved with the regular application of coconut oil.

PLAITING

You should first make sure it is squeaky clean – dividing it into sections only turns the spotlight on its condition at the roots, and any speck of dandruff, dirt or grease will ruin the whole effect. Ask around to find out how many plaits you should have and you'll be told anything from seven to twenty. Seven or nine – plus the forelock – used to be traditional for hunters, but nowadays a total of ten to twelve seems more usual, and sometimes even more appear in the name of refinement amongst the daintier show ponies. How many, seems to depend on what would look best to enhance or disguise the conformation on your particular pony. A pony with a thick neck could benefit from more and smaller plaits secured down the offside so that they don't show above the top line. A pony with a thin neck could be improved with a smaller number but plaited upwards and fixed to the top line to give the illusion of more neck.

I am a great admirer of other people's perfect plaits at shows, and although a believer in the old adage that practice makes perfect, I seem to have been practising for an awfully long time and perfect is not a word anybody has ever used about mine.... However, they are a far cry from the row of ping-pong balls which Abbey wore with panache on her first outing, so hope springs eternal.

Forward planning are the key words. Check that you have a comb, plaiting bands, a reel of thread or wool, several blunt plaiting needles and a bucket of water and a sponge. Once you have decided on the size of plait for the pony, you can save time by cutting a nylon comb to the exact measurement and using it as a template. You can also cut corners by using rubber bands for extra speed, but matching thread or wool is correct for showing classes, although you can cheat and use a rubber band to secure the end and hide it as you roll up and secure with thread. In any case, you will find rubber bands useful to hold the sections in place before you start.

Show-jumping ponies don't need to be plaited, but if you want to do so for the sake of neatness, then the hair around the withers should be left so that it doesn't pull and cause discomfort as the pony stretches his neck. Part-thoroughbred riding ponies with silky, designer manes make this job quite enjoyable once you've got the hang of it, but native ponies with hair like candy-floss can be trickier to cope with. My special aids are lots of water to dampen down, and a dollop of setting gel at the top of the plait to control the 'Afro' look. Egg white is a recognised setting agent for manes, and presumably promotes healthy hair at the same time, but having tried this once and ending up with one very slimy pony and up to my elbows in eggs, I didn't think the end result was any better than my proprietary tube of gel. Either I wasn't doing it properly – surely it wasn't meant to be cooked – or it stems from the olden days before gels and hair lacquers were invented.

Pulling and plaiting a tail

PULLING

Some ponies can react quite violently to tail pulling, so this is best done after exercise when the pores are open and the skin is warm. Starting at the top and gradually tapering off when you get half-way down the dock, pull a few hairs each side, checking for a balanced outline as you go. As with mane pulling, however good natured and co-operative your pony, do only a little each day until you achieve a neat result without any discomfort to him. Dampen down and bandage the tail after each session.

Pulling is rough on the hands unless you cover your fingers with sticking plaster, so wrapping a few hairs at a time around a tail comb before giving a sharp tug is recommended. Otherwise don't let metal combs or dandy brushes near fine tails, just gently separate any tangles with your fingers.

PLAITING

Many owners prefer to see a tail free and flowing, racehorse style, but where a pony enters Mountain and Moorland, Arab or Palomino Breed classes, the tail should look as natural as possible, so plaiting is often the only way for a neat turn-out. In ridden showing classes, a pulled tail is preferable.

Plaiting takes lots of practice to get a uniform tight finish, so don't wait until the morning of a show for your first attempt. Make sure the tail is clean and thoroughly tangle free. Wet the hair or run a little baby oil through it on your fingers to give it a sheen and to prevent the fly-away look. From the top, take thinnish sections in each hand from either side and proceed as illustrated.

With the left hand take a piece of the same thickness from the centre. This gives you the required three strands for plaiting. You proceed with the strand in your right hand under the centre section, plait it, and pull it tight. Now take another strand from the left side, into the centre, plait under, pull tight. Using your left thumb to keep the plait in place, add a section from the right and repeat this procedure until you get to the end of the dock. You then make a long plait from the three strands, securing the end with an elastic band or a needle and thread. Loop it up and tuck it under the first section, sewing it into

place to anchor it there, and over-sewing down the sides of the loop so it lies flat.

The length of the tail may vary a little depending on the pony's natural tail carriage, but not more than three or four inches below the hock should still be long enough even if he carries himself quite high. Use very sharp scissors to cut the tail in a perfectly straight line across.

If a pony rubs his tail, ask your vet to check it for sweet itch, which is caused by midge bites (see page 69), and can turn any tail into a lavatory brush overnight. But itchiness can also be caused by too rich a diet, by lice or worms ... so finding the cause is necessary before effecting a cure.

Dusty arrived to live with us with the proverbial rat's tail caused by years of over-pulling. We used a coconut oil conditioner and a soft bristle hairbrush (donated by Sam from her dressing table), and slowly nursed it back into quite a handsome ponytail. We plait the bottom to make it look fuller when brushed through at shows.

TAIL BANDAGES

There's a not very nice tailpiece to add of the gory catalogue of disasters which has emerged from the seemingly harmless, indeed well-meaning habit of protecting a tail with bandages. One such disaster, much publicised at the time, concerned a Guernsey man who bought a pony for his daughter in England and arranged for it to be shipped to the Island. This of course meant an eight-hour sea passage, to say nothing of the additional time spent *en route* to the ferry. When the elasticated tail bandage was removed, the shocked new owners discovered that lack of circulation over this prolonged period had almost caused the tail to drop off. Despite every attempt made to save it with careful nursing over several weeks, in the end it had to be surgically docked.

Some of the elasticated bandages on sale – bearing famous and apparently reputable brand names – stretch to twice their normal length, and leaving them on for long periods will naturally interfere with the blood supply. The manufacturers understandably defend their products by saying that anybody who knows anything at all about equines should use common sense when applying bandages.

So what are the safety procedures? Well first of all the British Horse Society's Welfare Department points out that wetting the tail hair, or worse still the bandage itself, causes an even greater restriction when

the material is dry, and wetting bandages is not a practice they recommend. Even dry non-stretch bandages shouldn't be left on for long
periods. Twisting bandages for a better non-slip grip, or tying the
tapes tighter than the bandage itself, can lead to trouble, and extra
precautions should be taken. Where a pony needs to be bandaged for
more than a couple of hours, for instance travelling long distance, use
a tail guard and a tail sleeve or stocking secured by a strap to the roller.
Remove the bandage by grasping the top with both hands and pulling
the whole thing off in one movement. Tail bandages should always be
washed thoroughly after use. This not only prolongs their active life
but avoids infection which may be caused by rubbing.

WHISKERS

A pony uses them as feelers, just as a cat does. Nobody would dream of shaving them off a prize exhibit in a cat show but in the showing world of ponies they seem to be regarded as superfluous and are usually removed by singeing, cutting with scissors or clipping. Unless you are keeping a mainly stabled pony for top class showing, think about this carefully because he may become disorientated by their removal. A tiny foal with crinkly whiskers bumps into everything. As he matures, his whiskers warn him of electric fences and the nearness of objects in the dark, and help him examine and select the food on offer. Also for cosmetic reasons, the downy hair inside the ears is often removed, and yet it too was put there for a purpose, to protect the delicate eardrums from dust and insects. You can always improve the general outline of the ears by holding the edges together and trimming round them with sharp scissors, especially the mutton-chop bits that stick out at the base.

FEATHERS

These provide natural protection throughout a cold, wet winter when they shouldn't be clipped right off the feet. Leave at least some hair on the fetlock to channel off water. Ponies with white socks and finer pink skin underneath are more likely than their darker friends to suffer from cracked heels etc. Light hooves are, in general, not as strong as dark.

Rugs

Nature provides a pony with effective insulation by growing a thick coat in winter to conserve heat. If he is to be wintered out and given only light work, you don't have to groom at all, except for the area where the saddle would cause sores if left dirty. The combination of grease from the glands under the skin surface and the layer of mud and dirt picked up from his instinctive habit of rolling on the ground, produces a matted weatherproof finish. This may not be so pleasing to the eye of the beholder, but it provides a most satisfactory protection from the elements and that's the important thing.

Any interference by grooming, necessitates stabling and rugs to compensate. A pony with a thick coat can't do any strenuous work, for the simple reason that the more active he is, the more he needs to sweat. With a thick coat he can't sweat properly and he'll quickly become damp and exhausted, running the risk of a chill. If he's to be left unrugged with just a little work from time to time, clipping the hair from underneath his chin and down the chest will help him sweat more easily. Otherwise he'll have to have his coat partially removed with any one of a variety of clips which depend on his use in winter, whether he's stabled all or part of the time and whether the body heat lost through clipping can be contained in rugs.

Stabled ponies turned out in the daytime are kept clean and warm with waterproof New Zealand rugs. The authentic design doesn't have a surcingle and you'll find such self-adjusting New Zealand rugs on the market which allow air to circulate freely and in which the pony can roll, buck and gallop around, without dislodging the rug. These are obviously preferable, and less likely to rub, than the ones held on with a mass of straps, but they are not so easily available.

The most available ones have breast buckles and hindleg straps plus a surcingle, which, unfortunately, far from keeping it in place, stops it from returning to the correct position once it's dislodged. This is why a lot of owners prefer to turn out ponies 'in the buff'. Ill-fitting rugs which work their way half off, rub patches of hair from the shoulders, press on the spine and let water trickle inside, are worse than having no rugs at all.

To fit a rug, lay it in a front-to-back direction so that it covers the coat from in front of the withers and fits snugly round the base of the neck and lies over the tail roots at the back. The edge of the rug mustn't sit on top of the withers or expose the shoulders. Fasten the breast buckle and check that the seam down the back lies in a straight line. Do up the surcingle. If the straps pass in between the hindlegs or each other and fasten onto the same side, you take the left strap at the front fastening, check it's not twisted and pass it between the hindlegs, to clip at the left back fastening. Now take the right front fastening, but as you pass it between the legs, link it through the left strap and then clip it on to the right back fastening. It's the linking which prevents rubbing by holding the strap away from the inside of the leg. So you can cross left front to right back, right front to left linking as you go, it doesn't seem to make any difference.

Remember an important habit for safety's sake is always to undo rugs from back to front. If by any chance the animal breaks loose, then the rug is less likely to cause an accident when still fastened in front as he takes off. If the front is undone the surcingle or back straps can easily become tangled around his quarters and back legs, causing him to panic and very likely to hurt himself in the struggle to get free.

When properly in place the straps of a New Zealand should be only just visible, not too high so as to irritate him or restrict his movement and not so low that he can catch in them as he rolls. Other rugs in the pony's wardrobe might include a jute stable rug, preferably lined, but you can always add extra warmth with a spare blanket when necessary, or a smarter quilted rug which is light, warm and easily washable.

A sweat rug is essential and acts on the same principle as a string vest under a shirt, allowing air to circulate to cool the pony down and dry off or retaining body heat to keep him from catching a chill. It isn't effective on its own.

A Summer sheet has many uses: on top of a sweat rug, under a heavier rug in winter – some ponies are sensitive to wool just as some humans are – and in the stable or travelling to shows when the weather's warm in summer to keep him clean and his coat lying flat. You can get smashing closer-woven cotton rugs which double as a summer sheet and sweat rug – expensive, but they look it, being ultra chic. They're practical and, in any case, you save part of the cost by not otherwise having two.

Woollen day rugs look smart and are of course warm but tend to be very expensive. However, a rummage round the second-hand goods in saddler's shops or at sales, often produces bargains in good quality rugs. A spare is always handy when one is being laundered or mended.

Stable and travelling rugs are held in place by a good fitting roller, or surcingle, and a fillet string which goes under the tail.

The most effective way to dry a soaking-wet pony is to cover him in sections of clean straw to make an undervest beneath a sweat rug. Dry the legs by giving them a brisk rub with loose straw or pieces of old towelling. Even if a pony is living out, he must be dried off and cooled down after working before being turned out into a field in bad weather, otherwise he runs a real risk of catching a chill. As a rough guide, you can test how cold he is by feeling the base of his ears. If they are cold to the touch, that's how he's feeling inside. As well as his ears, his loins are particularly sensitive to temperature change, which is why thoughtful pony parents place a quarter rug behind the saddle if he's standing around for long periods, at shows for example.

Once a pony comes back into work after a holiday where he's been turned out rugless, putting him into a New Zealand rug will flatten the winter coat and help him to lose it more quickly. Grooming, good feeding and gradually increased exercise then speed up the process.

Holidays

I think the idea of turning ponies out with only their natural coats and a field shelter for protection worries a lot of pony owners, and most children's ponies are turned out for their annual break when the summer is over.

Our ponies have about two months complete rest when they are turned out in the autumn, and they then come back in time for Christmas. I must admit it bothered me initially that an old pony like Dusty might not stand up to the worst of the British weather, so for the first couple of winters he wore a New Zealand rug by day and came in as night was falling. Two winters ago, with our vet's encouragement, I roughed him off in October, reducing, along with the exercise, the amount of concentrated food and feeding more hay and grass. We stopped grooming him, except under his roller and saddle, and reduced

Margaret Avison (age 11)

84

his stable clothes until he stayed in rugless but with the top of the door open, and then finally as his coat was becoming nicely woolly we chose a mild night to leave him out. From then on he was left with his natural coat and plenty of food to see him through the weeks that went into winter.

When early in January he came back from his farm holiday he was like a little white bull, fat and incredibly hairy, with woolly leg warmers and so many tufts round his head and feet he appeared to be wearing a hat and yeti boots. The following summer he produced the shiniest coat and the best weight and condition we'd ever seen him in. This 'return to the wilds' system may not work for every pony, but it seems to do wonders physically and mentally for our boys.

7 IN THE SADDLE

Pauline Sim (age 15)

When Michael, our son, was five years old we took him for his first riding lesson. The toothy instructress gave him the third degree. 'Bashful I think,' she finally pronounced, referring to the bomb-proof pony, which was a favourite with miniature riders, and not our son's character. One of her minions went off to waken the pony and she turned her attention to the new pupil's parents. 'Father looks the athletic type – nice long arms, good back – and what about you?' A quick scan of my person was obviously less pleasing. She said she was greatly encouraged when the child came from the right stock. Apparently she'd had the most dreadful shapes turning up and it was virtually impossible to do anything for them.

The lesson had both parents spellbound with the wonder of this small person managing to sit for a whole half-hour on a fat, hairy slug at a walk. Judging from Michael's expression, however, it was nearly as exciting as watching paint dry.

Later, over a drink to celebrate the first steps to Wembley, we recalled the lady's remarks about breeding blue-blooded equestrian children. We came to the conclusion that our five year old, ready for schooling, and our yearling, who would probably make 15 hands, were in with a chance as horsy children, being by a three-quarters thoroughbred out of a highland mare. Supposing, by a simple twist of fate, I'd gone off with Jack the registered Dartmoor after that bachelor-girl fling in Torquay? Our combined native blood might have produced a child with a near hump-back that would ruin the line of the showing jacket.

Now it was clear to us why the offspring of neighbours made it to the ribbons at every show. Richard was a well-built Irish hunter type with the natural conformation and pedigree required to register with a cavalry regiment. He'd been standing at the mess bar when this potential brood mare, Carol, was run up to him by her chestnut mother. Noting the good quality tack from Gucci and Hermès, and having his father's eye for a promising filly, Richard's interest quickly transferred to his oats. By Richard, out of the first-ridden Carol, came Sonia, a leggy and spirited youngster who would stop at nothing....

It's a game for any amount of players. The first one to produce a mother-in-law with the teeth *and* legs of the Derby winner, claims the prize.

Riding lessons

Before embarking on advice about riding instructors, I've done this round robin of all the people who teach Sam, just to confirm what I always suspected, that none of them holds a recognised instructor's certificate. Yet one of them, an ex-show jumper, runs a highly rated establishment in Oxfordshire where top international riders (the majority also without recognised teaching qualifications despite their success in equestrian sports) hold 'clinics' for adults and children alike.

At one of these clinics recently Harvey Smith raised a few eyebrows by dismissing lessons for children as over-rated, arguing that those who are keen will develop a natural ability, and learn the rest, as with many things, by getting on and doing it, and putting in as much practice as possible.

Scores of magazines and text books for pony and rider will advise you to choose a school which has the approval of a recognised society, with instructors who hold at least one of the BHS teaching certificates, which range from the assistant instructor grade, BHSAI, to the Fellowship grade, FBHS. This is the safest course when catering for the requirements of a wide and varied readership, and any other advice would be irresponsible. It is the obvious course to take if you don't have your own pony and are learning to ride from scratch. BUT it should always be remembered that these approved establishments are a recommendation and not a guarantee that you will get only the best at all times, and the inspectors who do the approving cannot watch over them constantly to make sure that nobody bends the rules.

For instance, a few years ago I had a bad fall from an Arab stallion in the Arabian Gulf which left me with a fractured pelvis and shattered nerves, so I was reluctant to get going again. It's common to a lot of women, I believe, that once they have children they won't take the same risks as they might have done before. A year or two went by before I ventured to enrol for a refresher course at a BHS approved establishment. An assistant instructress, wearing chipped, red nail varnish, gold hoop earrings, and with hair that looked as though it hadn't been washed since it had been on her head, conducted a class lesson sitting on a stool in the centre of the indoor arena.

Her voice, scarcely audible due to the combined lack of confidence and enthusiasm, and the fact that she was chewing gum, was drowned by the hacking cough of the tired old horse I'd been given to ride. Later at the office paying the money for this farce, I was asked when I would like to book again. I said I wouldn't like to, and gave the reasons. Some time afterwards at home, I got a call from the proprietor offering to take me herself, pointing out, quite rightly perhaps, that the class I attended was not suited to my individual requirements as I was not a true novice but someone who just needed a confidence-building refresher course. Quite so. Greatly cheered, I went back.

At the first lesson I was asked to lunge the horse for twenty minutes before getting on her. Not a bad idea in the circumstances as she was straight out of her stable and a bit exuberant for a riding school mount, plus I could see how she moved, and generally assess the animal. So far so good. Then the instructress took over, holding the horse, me and the lunge rein in one hand and her ten-month-old baby in the other arm.

Subsequent lessons were spent lungeing (by me) for half this expensive time to settle and assess the horse. On my last visit, she suddenly reared up then fell over in a panic at an unexpected noise outside the arena. Fortunately, a nice working pupil came to my assistance (my highly qualified instructor was making a phone call, and no one else was around), and so we calmed the mare down and put her away. That morning, the pupil had also decided he'd had enough and was going home to his mummy. On the way to the station, this disillusioned chap told me of his plans to become a hairdresser, pointing out that if conditions for soldiers were only half as spartan or unrewarding as they are for a good many working pupils in approved establishments, it would be the end of recruitment for the army.

Where money is scarce for lessons, even guidance from the parent with the aid of a good book (see Recommended Reading on page 141) is better than a child struggling along entirely alone. Go by reputation and results when finding someone to teach your child. Occasional lessons are good for the pony too as both can easily slip into bad habits. Freelance instructors are often less expensive if you are providing your own pony and paddock for the lesson. Just as there are horses for courses, so are there teachers. Some are wonderful at flat work, others specialise in improving the jumping ability of pony and rider. Occasionally an instructor will be brilliant at both, the theory being that when the flat work is right an improvement in jumping will surely follow.

A novice show-jumping pony who hasn't seen a coloured pole is going to be handicapped where the facilities on offer are a couple of oil drums, a rustic pole and a straw bale. And a teacher who was once the top event rider in the county may not be the best person to teach you and your four year old how to win the lead rein class. It's wrong to assume that because someone wins prizes they are qualified to help you. A good teacher has that extra quality of communication. The holder of an Olympic gold medal might be no good at all for the job.

My personal view is that riding school lessons for any child under seven are a waste of money. Some experts say that not until the age of ten do children have enough length of leg and sufficient maturity to get the best out of professional instruction.

Before they begin it's a good idea to let them watch a class where others ride. Once you've selected the school to suit your needs, go along and ask if you may come with your child to watch, and ask a few questions which will indicate you know a thing or two. Some instructors hate parents to watch a child's lessons, so it's always best to ask about this first, and also many children are intimidated by the presence of an over-ambitious parent and will often progress faster without an audience. On the other hand, freelance, Pony Club, or riding school instructors do not run baby-sitting services, and to drop a child off early in the day and leave her hanging around long after the instruction is over, is just inconsiderate.

You will note on arrival if the yard is swept clean and buckets and tools kept tidy. Say you'd like to see where the lessons will take place. Larger centres usually have an indoor school, show jumps and a cross-country course. At the very least, there should be a safely enclosed well-drained paddock. Find out how many riders are going to be in the class because a solo instructor can't be watching everywhere at once and is unlikely to be able to cope with a large herd of ponies stampeding.

Make sure you get satisfactory answers. Once when questions were invited at an early riding lesson, I asked, 'How do I stop him if he runs off with me?' and, 'How do I fall off?' and merely got a withering look in reply. Years later, when I spent weeks hobbling around on crutches after falling off a bolting pony, I had time to wish I'd persevered.

Children as well as adults fear ridicule if they display their ignorance, which is quite wrong. I know someone who gave up a rather dishy boyfriend when he asked how long it would be before her six-year-old 12.2 hands Dartmoor grew into a horse. And yet he grew up to be a Master of Foxhounds. We all make mistakes.

A child has to learn the hard way what adults find out all too easily, namely that the person you choose to show off to, will turn out to be better than you in the end. Those who tell you how well they ride, can't. Like the best lovers, the best riders talk the least.

Up-to-date lists of riding establishments approved by the inspectors are available from the following:

The *British Horse Society*, British Equestrian Centre, Stoneleigh, Kenilworth, Warwickshire CV8 2LR.

The Association of British Riding Schools, 54 High Street, Penzance, Cornwall TR18 2HY

For approved Trekking and Riding Holiday Centres contact
Ponies of Britain, Ascot Racecourse, Berkshire SL5 7JN

The BHS Blue Plaque, at the time of writing, is displayed on more than 500 riding establishments throughout Britain. This does not mean that they are necessarily the best in the country. The BHS scheme applies only to those who volunteer to join. Because of this voluntary process the Society does not, indeed would not, claim that all establishments who do not apply are inadequately run. To start in business as a riding school you must have a licence. The Trading Establishments Act Committee works closely with the authorities responsible for issuing licences. They make regular visits to every licensed establishment in the UK other than those on the approved list.

What the BHS's approval signifies to those who voluntarily seek inspection, is that in their opinion the school provides good riding instruction and an accurate description of the facilities. They also provide, among other things, an advisory service to proprietors who qualify, and make available at discount rates third party insurance for the protection of both client and proprietor.

Although an appointment is made at a time mutually agreeable to proprietor and inspector on the initial visit, once the establishment is listed as an approved school it can be visited at any time without warning, and letters and calls from members of the public concerning complaints will be investigated quickly.

The Pony Club

The Pony Club is coming up to its sixtieth birthday, and there must be many mothers and grandmothers of today's young riders who have family albums filled with photographs of perhaps the more common types of ponies than the average Pony Club mounts of the 1980s. Former members include Princess Anne, Captain Mark Phillips, Virginia Leng, David Broome, Mandy Rice-Davies and the Duchess of York. They all had their formative lessons at rallies or camp, gained places in teams for the Prince Philip Cup, Tetrathlon, showjumping, horse trials, passed the tests and wore the badges with pride.

There are 366 branches of the Pony Club in Britain, and a good thing too. Pony-mad children wherever they live can benefit from free instruction and receive practical help and advice concerning all aspects of riding and pony welfare. The nice thing is that you don't have to have your own pony to join, and help is always willingly given to find a suitable pony to loan for a week at camp, or perhaps to exercise from its present home.

The standard is high at Pony Club competitions, mainly because the

Julie Clarke (age 11)

upper age limit is twenty, which some say is too high, arguing that the Pony Club nowadays has little room for run-of-the-mill ponies who haven't cost thousands of pounds. However, that criticism may be a little unfair in view of the fact that the Pony Club does try to cater for riders and ponies at all levels. How can you stop some children having better ponies than others? A point to remember is that money alone doesn't guarantee success. It certainly helps to have a smashing talented pony, but he's no good – whatever his price tag – without a child who is equally capable of getting the best out of him.

To find out more about the Pony Club, write to The National Equestrian Centre, Stoneleigh, Kenilworth, Warwickshire CV8 2LR. They'll put you in touch with the District Commissioner of your local branch.

Road Safety

I would urge parents of young children to enrol them as Pony Club members if only for the Riding and Road Safety test for children of ten years and upwards, who will at some stage take a pony out on the roads unaccompanied. The test is of a high standard, with members of the police force in attendance. Children must satisfy the examiners by answering questions correctly and giving a practical demonstration in traffic that they understand the full implications of being in charge of a pony in hazardous road conditions.

The test starts with an examination of the rider's clothing, particularly the hat to make sure it is a well-fitting British Standards Institution Number 4472 model which has been compulsory wear at all Pony Club events since January 1986. The pony's shoes are inspected to make sure they are neither worn nor loose, and tack is given an over-all appraisal to check the stitching is safe, stirrup irons are the right size etc. In view of the fact that statistics show the likelihood of one in seven riders being involved in road accidents, which occur on average at the rate of eight per day, passing a road test makes sense. They are not only run by the Pony Club. Riding clubs also arrange them and your local County Road Safety Officer can advise you about others.

The BHS booklet *Riding and Road Safety* and their *Manual of Horsemanship*, in addition to *The Highway Code*, are recommended reading.

Here are some of the questions children have been asked in recent Pony Club tests.

1 On which side of the road should you lead a pony?
2 Describe your position in relation to the pony and traffic when leading the pony.
3 Should you lead a pony in a headcollar on roads?
4 Describe your position in traffic if riding round a roundabout.
5 If riding down a steep hill, what should you do?
6 Do riders and motorists follow the same procedure for turning right on a road?

Answers
1 On the left side, as you would if riding him.
2 In between the pony and the traffic.
3 No, it is illegal and unsafe, you have more control with a bridle.
4 You must always keep to the inside of a roundabout.
5 Get off and lead the pony.
6 No. A motorist may signal and pull into the centre of the road until it is safe for him to cross to the other side. A rider remains on the nearside, gives the signal, and when all is clear moves to the nearside of the turn in.

Fluorescent arm and leg bands, tabards, stirrup lights and other road safety aids can be obtained from the British Horse Society's headquarters, or from tack shops or by mail order through pony magazines, or enquire at your local road safety office.

For bad weather conditions ask your blacksmith to fit road studs, and always smear hoof grease (or even lard) to the underside of the pony's feet to stop snow or ice forming a ball underneath him, all of which are sensible precautions.

A nervous pony and child are best introduced to road work in the company of an adult on a 'schoolmaster' horse or pony. Bomb-proof ponies are hard to find. Some ponies who are apparently bomb-proof when first tried out, prove to be specialists in selected spooky things. Dusty is a pig specialist – a throw-back to when his ancestors were chased by wild boar. Abbey cares not two hoots about pigs, but we have learned to keep her at a distance from pony traps. Specialist ponies may often be cured of such little quirks by patient training to teach them that the spook in question means no harm. I do not intend to buy Dusty his very own piglet for Christmas, or to train Abbey to pull a trap, but if I did, I might well eliminate their distrust of them by

Jane Simmons (age 10)

constant repetition until the 'monster' becomes part of the scenery. Less discriminating ponies are looking for distractions the whole time, so by and large our boys are good in traffic, and quite fearless when they have Gay as a security blanket.

8 SHOW BUSINESS

Stephanie Cooke (age 16)

It's nine o'clock in the morning, and we got up four and a half hours ago. Rule Number 1: However early you start, it's never early enough.

Despite much poring over maps the night before, I managed to take the wrong turning and so we have cut it a little fine. How deserted country villages are in the early bright when you are looking for someone to tell you where you are. Normally everyone I ask for directions pretends to be deaf or potty, so today at least we've been lucky in finding a knight in shining armour – disguised as a decorator in a boiler suit – who rose to the challenge and drew me an idiot's guide to the showground. 'You can't miss it!' I know it's just a friendly conversational phrase like 'Have a nice day,' but it can grate when, quite clearly, you can.

However, half an hour later we limp into the showground feeling like Columbus when he discovered America. This venue has a novel twist for welcoming horseboxes. Great ramps have been placed at frequent intervals with muddy rutted tracks in between. The convoy humps and shudders, gets stuck, then lurches free to the strains of groundsmen offering advice and drivers returning it. Horses and ponies clatter around their boxes, and there are fearful sounds as though somebody has finally panicked and fallen over.

Rule 2: The secret of doing anything with ponies is to avoid upsetting them. We let down the ramp. They are not upset, merely as relieved as we are that we made it at last. Open flask of hot coffee and take a deep breath as the steam defrosts my nose. Remember our decorator friend who asked, 'Why do riders risk broken necks over fences, and terminal pneumonia in pouring rain, to win a few pounds, when snooker players can get thousands for pushing a ball around a table whilst enjoying a fag and a pint in a nice warm room?' Unbelievable how daft some people are. No time for riddles. Right lads, where shall we start?

What well-dressed riders wear

A *hat* is the first priority for any riding activity. If you are a Pony Club member, you are made to wear a BSI 4472 skull cap, compulsory in all Pony Club competitions, including dressage, so let this be your guide. Beware of second-hand hats, which might have been dropped on a hard surface and cracked, otherwise second-hand bargains exist in all the main items of clothing for riders. At shows, a hunting cap looks smartest

in show-pony classes, and it's correct to sew the tail ribbons up underneath the brim at the back. Regular showing jockeys have a crash hat
for jumping and a hunting cap for showing phases on the flat. It isn't
wrong to wear a crash hat with its chin strap for extra protection for
both phases in Working Hunter Pony, and it's obviously safer, it's just
that the hunting cap adds to the elegance of a turn out. Children are
hung upside down before the cap is purchased, to ensure a good fit
should they be upended when riding. Choose a colour to tone in with
your jacket. Velvet hats fade rapidly in sunlight, but an occasional
wipe over with a cloth wrung out in a vinegar and water solution, or
steamed over a kettle, will revive colour and pile considerably.

Jackets. A good fit is essential, however tempting it may be to get
one that you'll grow into, or to hang on to one you think brings you
luck, long after it's outgrown. You will always recover some of the cost
of a good jacket that has been well looked after, when you trade it in
for another one. It shouldn't show the cantle of the saddle or the top
of your gloves. For showing classes, plain coloured jackets, navy, brown
or black are worn, often with a velvet collar, but a tweed jacket is
preferable in Working Hunter classes, although navy or black are
permissible. As with the tack, in this class a rider's clothes should be
workmanlike. Don't wear flowers in your buttonhole.

Shirts. With jackets, for all showing classes and Pony Club events
stick to plain light colours, cream, white, pale blue or pale lemon.

Jodhpurs. Dark-coloured ones are out except for informal riding. Better
to invest in cream, beige or pale lemon. Again, they must really fit to
have a nice line. Stop the bottoms riding up and showing off socks by
sewing black or brown elastic (depending on the boot colour) into the
inside of jodhpurs to hold them in place when pulled over the boot.

Jodhpur boots. Brown is the most popular colour and looks nicer
than black, although black is also acceptable. Long boots are not,
unless on older children in Working Hunter classes, when good-quality
leather hunting boots are obviously best – straps should fasten at the
front with the tongue facing outwards. If you only have rubber boots,
use ordinary boot polish to give them a fake-leather shine.

Accessories. Muted colours are best to tone in with main items, except
in lead-rein classes where judges wear sunglasses to protect them from
the glare of brightly coloured hair ribbons, browbands etc., – from the
tinies upwards, colours tend to be more subdued. A showing cane
(maximum length 30″) covered in the same coloured leather as your

boots – in Working Hunter and of course show jumping, other types are used. A pair of leather gloves look smart in the show ring, but become mushy in wet weather and too slippery to get a grip, so the kind with non-slip palms are best especially where required for jumping. To complete a smart turn out, girls wear hair nets. Pull them forward at the forehead, and twist the surplus around before pushing it under at the front. Once the hat is in the right position – straight, with the brim pointing forwards, not up – it will keep the hair net in place. Long bootlaces in colours to tone with the jacket are the answer for quick threading through number cards at shows. Cutting the corners off the cards, so they sit snugly into the back, completes the neat and elegant over-all picture.

Back support. Strongly recommended where jumping across fixed fences and other more-hazardous-than-usual pursuits. Children do not on the whole choose to wear these without encouragement, but they make sound common sense.

What the well-dressed pony is wearing

Saddle. Riders who compete in ridden show-pony classes, as opposed to Working Hunter and showjumping classes, have different requirements. A show saddle has a straight seat with straight cut flaps which show off a pony's good shoulders and top line. Sam uses the same saddle for both our ponies at shows, but because Abbey has a shorter back, Sam has a 'point' strap fitted in front of the other girth straps, which stops the saddle slipping back. A strip of foam rubber or a piece of dimpled rubber inside the girth will temporarily stop any movement. Always take the advice of a proper saddler, not a tack-shop assistant, when investing in this costly item. If you are buying new, and cannot take the pony to the saddler, ask him to come out to measure and bring a few saddles to try. You must be able to see daylight along the pony's backbone, even when the rider is in the saddle. If buying second hand, explain what purpose you have in mind – show jumping, hacking, showing, or a bit of everything. Measure across the wither with the aid of a wire coat hanger straightened out first then moulded to the pony's shape, in the same way as a dentist takes an impression of a mouth before fitting a brace. Give the pony's height and say whether he's long short or medium in the back. Ask if you may return the saddle if it's an incorrect fit, and get somebody who can tell if it's comfortable for the pony to give their approval before you decide. Is it comfortable for you? Buy it.

Girths for show ponies are made of white tubular webbing, but save them for those special days – they are murder to keep clean. Working Hunter ponies wear leather (expensive but good-looking), lampwick (soft strong fabric reinforced with leather at the buckle end), and nylon cord or polyester mix girths which come in a variety of colours. Avoid the flash blues, reds and greens if you intend it to double as a hacking and showing girth.

The same applies to coloured *numnahs*. A numnah gives greater protection to the pony's back, unless badly fitted under the saddle, in which case you would be better off without it. Always take your saddle with you to buy a numnah of the correct size. Push the numnah up into the channel of the saddle leaving a clear space from the wither along the backbone and make sure that it doesn't pull down again as you put it on. Unless it has an even border, clear of the rim of the saddle all round, it will cause friction and subsequently sores. Most ponies at shows don't wear numnahs, and if a saddle is correctly fitted you may not need one at all. If a saddle isn't correctly fitted, change it or have it re-stuffed, but don't use a numnah as corrective padding. Old ponies like Dusty are often glad of a numnah. He wears one for most things, and always when hacking, cross country or jumping. But beware sheepskin numnahs in summer or when they are doing strenuous work, they get very hot and sweaty. In any case, numnahs should be kept clean with regular washing to remove sweat and grease marks, they also collect the dirt which would otherwise build up on the saddle.

As well as a well-stitched strong girth (the best buckles have a top groove which prevents the tongue slipping) an elasticated surcingle round the saddle is a standard precaution for cross-country events.

Bridles. Double bridles are worn by show ponies above the lead-rein and first-ridden classes, but they require a practised hand to use them correctly. In Working Hunter Pony classes, double bridles, although not necessary, are often used to diminish the size of a chunky head in the same way that bigger saddles are used to good effect to disguise too long a back. Velvet coloured browbands can be worn at shows, except in Working Hunters where plain, business-like tack (broader flat nosebands, for instance) is used. All browbands, nosebands and bits must fit correctly without pinching the ears, the mouth or the nose and restricting the breathing.

Martingales. Used where the pony is carrying his head high, evading the bit and can't be properly controlled. Take specialist advice with these, and also if you think a change of bit is necessary – you can spend a fortune collecting accessories which are useless for your particular pony.

There are basically two types of martingale: standing and running. Standing martingales exert pressure on the nose and should never be used with a drop noseband or when jumping. A running martingale presses on the mouth and must be used in conjunction with leather or rubber rein stops. The combination of a double bridle and a running martingale is often seen at shows despite the fact that most child riders haven't the necessary experience to use them properly.

Keep tack clean by regularly using a high-quality saddle soap for basic suppleness, but finish off for shows with a rub with beeswax furniture polish or wax shoe polish for a high gloss on all but the parts where your grip would be affected, upper reins, saddle seat etc. A new saddle needs several applications of neatsfoot oil, and occasional protection with a leather dressing at intervals, especially to protect against rain or extra dryness in summer.

Beauty secrets of the star ponies

Perfectionists will shriek that there is no substitute for thorough daily grooming combined with a good diet. The only extras they might use to achieve a glossy coat would be cod-liver oil, milk pellets or boiled linseed for an inner shine, coupled with a daily 'strapping' session to tone up circulation. Cleanliness is essential for good skin tone, as any Miss World contestant will tell you, and beautiful ponies who parade in front of audiences have fairly similar requirements and professional tricks to enhance their natural beauty.

Shampoo. There are many good proprietary brands on the market. Bear in mind that any shampoo removes natural oils as well as grease. Some coats shine better without them, hence artificial shiners are applied afterwards to counteract the dullness after a bath. Hexocil is the recommended medical shampoo for ponies with scurvy and dandruffy

Zena Galvin (age 13)

coats, a condition which often arises through poor diet or inefficient grooming. It's only available through your vet or at chemists.

Conditioner. Coconut oil is magic on moth-eaten manes and tails to encourage growth and repair brittle hair. It's not for use on show days – it looks greasy and attracts dirt like a magnet, but it's fine on his day off – there's a price to be paid for everything ... like the cost of a dollop of your own conditioner on a mane or tail. Where you're not plaiting, conditioner helps a newly pulled mane to lie flat, and stops that flyaway look in tails. Spray on conditioners, like Show Sheen and Canter, leave ponies as shiny and slippery as miniature skating rinks but also protect the hair against dirt for several days after the application. Stockings, or tights cut up, are useful for popping over ponies' tails and securing with a tail bandage to keep hair clean.

Electrical aids to grooming come in the form of groomers and clippers. The bigger saddlers often have re-conditioned ones. We use dog clippers on heels and chins, but with caution, it's easy to get carried away and leave harsh lines where stubble grows. It's best done well before a show day, rather than have an entry who looks like he's been caught in a lawnmower.

Stencils are used to make easy work of quarter marks. Sold in plastic sheets of squares, diamonds etc., stencils are used in conjunction with a small comb, and the hair is then set with a dash of hair lacquer. They are not usually seen in Working Hunter classes.

Sharks teeth are marks made by changing hair direction on the quarters, using a brush.

BEST BUYS FOR THE MAKE-UP BOX

Vaseline, olive oil or liquid paraffin highlight the hairless areas, the eyes, nose, dock, and are often better than hoof oil on ponies with light-coloured feet.

An old silk square or a chamois leather makes a good final polisher.

Hoof Oil. Clean cooking oil does as a standby. The best proper 'hoof oil' I've found is by Valley Sales of Bledlow Ridge, Buckinghamshire, who sell through mail order or from stands at shows. Each application lasts three days. Hoof *polishes* are very popular now, but my blacksmith says they dry out the feet. They come in clear or black, the black looks like gloss paint as if the pony has patent-leather tap shoes. Some show producers use grate black, I am told.

Chalk. In block or powder form for white areas on legs or faces. *Shoe whitener*, the kind with an inbuilt sponge applicator, is also used for brightening white areas. The black variety darkens hocks.

Setting gel is marvellous stuff for last-minute flyaway bits on manes and tail tops.

Hair lacquer. Apply as you plait.

Sun-tan oil for ponies with pink or sensitive skins on sunny days. Après Sun, zinc or calamine lotion can also be used.

Baby or olive oil. Rub into palms of hands, then lightly pass over knees, hocks and through tail.

Brylcreem is another popular highlighter for knees, hocks and unruly hair. Apply on a cloth to avoid streaks or patches.

Reckits Blue. Add to the final rinsing water for whiter-than-white socks and for taking the yellow tinge out of grey tails.

Bargain baskets at larger chemists and department stores contain El Cheapo sludge-coloured eye shadows or foundations to accentuate the positive areas or to tone in with natural colour to disguise stray white hairs or other blemishes in the coat.

Other common ploys to distract the attention from defects may include shoes made from light aluminium alloy to increase floating action. These must be put on just prior to showing – it's the sudden change from heavier-weight shoes which exaggerates the movement, like changing out of muddy wellies to go dancing in satin pumps.

Blacksmiths will tell you a whole range of dodges like making pigeon-toes appear square by off-setting the clips, and boarding up cracks and chips in hooves with proprietary brands of filler from DIY shops. They will advise against new road-weight shoes prior to show days in case the pony drags his feet.

See page 76 for how to create an optical illusion with manes.

LISTS

Well worth making. Put them on postcards in the tack room, transport, jacket pocket and anywhere else for convenient last-minute checks. A list can go on for ever, but here are some of the basics:

Saddle
Bridle
Girth
Stirrup leathers and irons
Spare browbands and bit if necessary
Numnah
Spare rug
Spare tail bandage
Jumping studs if necessary
Martingale if necessary
Lunge Line if necessary for loading, or exercising.
Hat or hats
Jacket
Rainwear/extra sweaters
Shirt
Tie
Gloves
Stick/sticks
Clothes brush, shoe shiner, cotton wool, tissues, hand towel
Large water container
Buckets
Haynet
Grooming box: body and water brushes, curry comb, hoof oil and brush, hoof pick, stable rubber, pony's make-up (baby oil, Brylcreem etc.)
Sewing kit – with emergency mane-plaiting bands, needles, scissors, wool or thread.

Spare buttonhole, hairnet, ribbon or bootlace for number.
MONEY, SCHEDULES, MAP AND DIRECTIONS OF HOW TO
GET THERE.

Your pony, unless you hack to a show, will be wearing a rug, surcingle,
tail guard/bandages, protective travelling boots, a headcollar and rope.
We find it's best if the rider dresses in the jodhpurs, shirt and tie before
leaving home and wears a track suit and rubber boots on top to keep
out the dirt until we get there. Get the pony worked in and then change
into the jacket and jodhpur boots, which can be given a last-minute
shine along with tack (use a shoe-shine pad) once she's in the saddle.

Ponies keep tail and leg bandages on until just before the class. You
must take care that the leg bandages are absolutely secure and won't
unwind while the pony is working. Ponies going over cross-country
courses have their bandages sewn for extra security, or bound with
electrical tape.

COB

For all I enjoy being Head of Wardrobe and Make-Up, Catering Super-
intendent, chief groom, navigator and driver, I don't feel the slightest
urge to become a competitor. If I didn't faint with nerves in the collecting ring, I would die of embarrassment once inside.

A few years ago, I regularly hacked out a Cob which belonged to
friends in Suffolk. He was kind and careful and built like a Chieftain
tank. The only danger to one's person was that you might walk forever
like John Wayne, unable to re-train your legs to meet in the middle.
We spent many hours together, Cob and I, he battling to control his
ever-increasing girth as I developed thighs like nutcrackers. We struck
a rapport based on a mutual regard for the other's limitations.

A few weeks before the Suffolk Show, Cob's owners said they were
planning to enter him and would I like the ride? He was an old hand
at Cob classes, I would just have to sit on and steer and leave the rest
to him. For some extraordinary reason I agreed, not wishing to appear
a bad sport, but dreading every single minute. Despite Cob's normally
genial disposition, I was sure that I would turn us both into gibbering
idiots on the day. However, I was almost as reluctant to renege on the
deal, so we set about our daily schooling sessions in earnest.

We pounded around fields of stubble and disused airfields in front of
imaginary judges, working out what I hoped was a reasonable show.

The fact that I couldn't get on him without a leg up, and that he had an unfortunate habit of breaking wind when changing gear from trot to canter, began to be the least of my worries. We would simply go in there and show them our beautifully synchronised performance. The euphoria was short lived.

One day as we ambled up a village street I nodded Good Morning to an old man chopping firewood in his garden. He straightend up, stood stock still and gazed in disbelief, eyes and mouth wide open as we passed. Then, 'Good Gawd,' he gasped, 'I aint seen an arse on an orse like that in Yurrs.'

I didn't stop to ask him to clarify the remark, to find out if he meant mine or Cob's. I decided there and then to risk being called a spoilsport and find someone else for the show who could cope with the critics better than I. I didn't feel the slightest twinge of regret standing at the ringside cheering Cob and his new, male, thirteen-stone rider on to victory.

It's not where you start, it's where you finish

For prospective pony-owning parents and children, a good time to study the market is early summer when the curtain goes up on a magical round of shows, from DIY affairs in somebody's back garden to the county variety, and others where only top-class affiliated ponies need apply.

Apart from at the very bottom end, where anybody can win so long as they are related to the organisers, or are pupils at the riding school sponsoring the event etc., or where the only available judges can't tell a quality pony from a pantomime horse, you'll find a good cross section of classes and competitors at local shows within a reasonable distance of your home each weekend throughout the summer.

If you already have a pony and have not yet ventured to show it, chances are you will want to do so before long, and weighing up the opposition before plunging in at the deep end will give you more confidence. Competitive pony followers buy the Spring editions of *Horse and Hound*, and other publications which contain the show lists, dates, venues and show secretaries' names and addresses, so that they can dispatch the annual mound of stamped-addressed envelopes for entry forms and schedules and plan the season accordingly. Really keen types will keep a show diary with a record of their winnings (if any), notes on whether last year's judge preferred pretty ponies to workman-like types and couldn't stand palominos at any price, whether the standard in the showjumping was poor and what height the jumps were, directions for finding the place etc. I know somebody who has more files than Scotland Yard, based on only four years of ponies. Studying form at these shop windows will help you understand more about conformation, movement, which type of pony suits the various classes, and what you need to have in the way of clothes and tack for a suitable turn out.

Try your hand at ringside judging to see if you can pick out potential winners. Watch in the ridden classes how children give an individual display of their pony's talents and their own ability to control him. You will also learn how the first impression a pony and his rider present to a judge can be all important, because a class can be won or lost from the moment competitors enter the ring.

It is here that you spot the pony with presence, who stands out from the crowd, and children who have learned a trick or two about ringcraft. They treat other competitors as though they have the plague, finding their own space in the ring, and, if caught in a cluster, circle round to find more room. They sort out any problems they may have with a pony playing up, behind the judge's back, and they don't make a big drama or pull angry faces if things go wrong as amateurs do. They know all the crafty ways of upstaging other ponies, keeping their distance from any pony of the same colour, putting their grey between a black and a bay to make it stand out. Once all the entries are in the ring, they walk (most likely on the right rein) until a steward gives the cue for them to trot on, and at another given signal show the paces at the canter.

Throughout all this, the judge has made notes about conformation, action and manners in the company of others, so he will proceed with the assistance of the steward to call them in, usually in a provisional order of preference. From this they will be moved up or down the line, depending on their solo performance and whether on closer inspection the pony confirms the judge's initial impression. So it is unlikely that a pony pulled in ninth in line will win, but he could get into the placings if there are rosettes for first to sixth and the forerunners let themselves down. Anything can happen at this stage. Ponies can buck and unseat riders or display other signs of bad temper and ill manners, which unfortunately is not always treated with the disrespect it deserves, especially if the animal is blessed with good looks.

Children new to showing become a trifle bewildered by what is expected in an individual display. A judge will either issue specific instructions, perhaps to cut down time in very big classes, or he may leave it to you. Some children embark on a marathon which bores everybody to tears, and the ones who make an impact are concise and polished in their performance. Unless a rider is the first to go, she can always study carefully the ones at the top of the line, pray that they know what they're doing, and copy them.

Work out a 'show' at home, using markers round a decent-sized flat area which doubles as a ring, and with a parent who doubles as a judge. Child and pony should walk smartly, not slouch, in a straight line and halt a few feet in front of the judge. Stand perfectly still and square with the pony using all four legs equally to support his weight. Use rein and leg contact to show off his head carriage and stop him fidgeting about whilst the judge has a good look all round him. It's therefore important to teach a pony to stand still, unruffled, even when strangers run hands down his legs, lift his tail and feet, look in his mouth or whatever. The judge will indicate when you should move off to start your demonstration of balance, control, good manners and good paces, because that is the object of the exercise which a lot of children don't understand.

Walk to the outside of the ring, and smoothly make a transition into a trot around the arena and in front of the judge, then just as smoothly begin to make a figure of eight at a canter, changing leg in the middle to show he is co-ordinated on both reins, and finally end with a stretch of gallop if it will be required in the classes you enter. Most people favour the far side of the ring at this point, but you must place yourself above all where the judge can see you without having to climb onto the steward's shoulders for a better view. In front of him, as much as you can, makes the best sense. A good low gallop is important in Working Hunter or Show Hunter class, less so in ridden pony classes, and not required at all in first ridden and lead-rein classes.

Unless asked to do so, you needn't rein back when you return to the judge, just show that the pony is quite relaxed by walking him up on a longish rein and again in a straight line as you halt. A big smile for the judge, and a girl bows her head and a boy removes his hat. That's it. 'Thank you, well done, unbelievable, don't ever come back,' or other suitable comments or nods from the judge will signal that he wants you to go back into line. Remember at a show to do this without getting in the way of the next competitor or disrupting other ponies as you take up your previous position in the line.

There are a variety of classes at other than the major shows where the utility pony treasured for his manners and ability rather than his looks, has every chance of winning something. You may be able to buy a pony, but you can't buy experience and such classes offer an excellent training ground for other events.

FAMILY PONY
In this class the pony should be suitable for all members of the family to ride (unless the class is divided into height sections), therefore it must be of a good build to take the weight of an adult and well-mannered enough to be safe in the hands of a novice child. Sometimes they are required to jump a small fence, or to illustrate that they will stand still whilst being mounted, a basic requirement of any ridden pony after all.

Although your entry may live in a muddy paddock with a field shelter, as a lot of genuine family ponies do, this shouldn't be made obvious to the judge. Too often Family Pony is interpreted as Scruffy Pony, but the line-up should still be one of well-groomed ponies and riders, so time taken to plait mane and tail will be appreciated. Although you will often see in this class all sorts of daring stunts performed by riders in an effort to prove their ponies are as docile as Cossacks, extraordinary routines like crawling about underneath him, hiding under his tail and twirling round in the saddle, shouldn't be necessary and are frowned upon by most judges.

HANDY PONY
This class is for nimble, unexcitable all rounders who can complete an obstacle course in the shortest possible time. No extra marks for his appearance, he can be crossed with a giraffe and still claim the top prize. A good sideline for gymkhana ponies.

VETERANS
This is one of my favourites, and is ideal training for children to become excellent poker players later on. Ponies from fifteen years of age, or however many it tells you in the schedule, are eligible, producing an enormous class which goes on well into the night. It can be like a

lottery with the best bet being on the eldest pony to win, which teaches children to tell fibs. In other classes you may be asked to produce a rider's birth certificate or a height certificate for a pony, but veteran entries are taken on trust. As the judge enquires as to the age of each one, sprightly looking animals who've just finished six rounds of showjumping are somehow transformed into quaint old souls.

'How old is this pony?' asks the judge.

'He's thirty-nine,' says the cherub on its back, and proceeds to tear up the ring at a flat-out gallop to show how carefully he's been preserved. Good for a laugh!

BEST RIDER

Usually divided into age groups, this is an interesting class since everyone rides better in their mind's eye than they do in the saddle. In theory you should be able to draw a line from the crown of the hat down through the rider's ear, shoulder and hip, and end at the back of the heel. Variations in this class will include the water-ski position – long reins, feet forward, leaning back in the saddle; the duck position – bottoms up, elbows out and flapping like wings; or the ballet position –

Helen Shortis (age 15)

feet turned out and toes pointing towards the ground, arms raised. Despite the class description, it's often a fun affair with competitors running into everyone they know.

CHILD'S FIRST PONY/FIRST RIDDEN

In theory looks shouldn't matter greatly in this class where safety is the primary consideration, but because it is one of the most popular showing classes, attracting large numbers, they most certainly do. Of course the same misnomer occurs in the Working Hunter section where ponies can't hunt because a child couldn't hold them and they might hurt themselves.

In a WHP Nursery Stakes class a few years ago, the cute little course included a stuffed fox sitting in front of brushwood, and four shocking-pink ducks on a water trough. Our Dusty sailed across the ducks – pink bobbing bath toys being, after all, an everyday feature of the British Hunting scene – but he just hesitated for long enough to incur faults at the foxy brush fence.

'Pity about that,' remarked the judge when they'd finished, 'he was going so well too.'

'Yes he was,' agreed Sam, 'but I should think he's terrified of foxes ... not that he's ever seen any, but he's scared stiff of pigs.' And on that recommendation for a bold hunter, they bowed and left the ring.

We've only had one truly genuine child's first pony, an 11.2 hands Welsh Mountain called Crackers. He was so good, a monkey could have ridden him. A few days before his one and only sojourn into a showing ring we had been out on a quiet leisurely hack and were approaching a busy road home, when without warning a cat fell from a tree, just as Sam and Crackers passed underneath. As the pussy scratched and scrambled down his neck, Crackers stopped, shook his head vigorously in irritation, then walked on. He'd seen it all over the years and with numerous children who'd had their first lessons, hunted and taken the Pony Club 'D' test on him.

That weekend, he stood down the bottom half of the line, whilst the dainty, porcelain-like winners received their awards. Everybody knew that the winning pony hadn't ever seen a road and got his only exercise on the end of a lunge line or inside a ring, but he looked angelic, moved beautifully and the judge was enthralled.

I was reminded of this one day when watching with another pony mother, Patsy Morton, a first ridden in progress, where one of the entries was a Crackers look-alike. Patsy had a special interest in the class, having owned this pony previously.

After the initial parade round, and as they were brought into line, it was obvious that this more Thelwell than thoroughbred pony had made very little impression, he was well down. As they were nearing the end of their individual showing, a muck spreader came clanking into the field behind and the stampede began. Small children screamed for parents, as empty ponies headed for the exit. Like the Tin Soldier, our little Welsh friend stood, totally untroubled, savouring his moment of glory, his pigtailed rider smugly watching.

To the judge's credit, this perfect example of a trusty first pony received his just reward. As he led the lap of honour, red ribbons flying, I remarked to Patsy, 'Well that cheers you up doesn't it, he's obviously bomb-proof.' 'Oh he is,' she said, and then added with a wink, 'he's also stone deaf, bless him.'

LEAD REIN

The class in which so many pony parents cut their first teeth. A child is a mere prop, dressed up like a dog's dinner to match the over-all décor. Mother, child and pony can wear the full make-up here. Mother's fashion accessories will include a smart brimmed hat, Sloane Ranger shoes and scarves in tones to match the pony, his flash browband, and the child's hair ribbons.

Until now you may have had the impression that riding children are only born to single mothers. However, for this class (and in Lead Gymkhana) the fathers appear. The object in dress for men is to look as much like the listening bank manager as possible. In order not to lose his red carnation and bowler hat, Father's movement as he runs before the judge is reminiscent of the tango or the palais glide, bringing tears of nostalgia to those who go back that far and occasional tears of laughter to those who've just arrived.

It's also very competitive now, and a far cry from my childhood visits to shows in Scotland, where heavily thatched Shetland ponies were secured at the end of bits of string. Caps, braces and tackety boots were standard gear for fathers who spent all morning competing for other trophies in whisky tasting. A Woodbine between the lips and a final hitch of their socks, and they were off. So too was many a child who fell unheeded from the saddle to wild cheers from the crowd.

GYMKHANA

Considered extremely naff by much of the showing alliance, as I discovered early on, when in my innocence at a show one morning I asked the mother of one of Sam's friends if they were staying on for the gymkhana. She looked as though I had just run a cheese slice down her nose. 'Ours is a show pony,' she answered between gritted teeth, 'and it teaches them such bad habits.'

Now I have to admit that when Abbey got the hang of gymkhana games and she found it all terribly exciting, we had to call a halt when introducing her to Working Hunter pony classes. Quick to catch on to habits good or bad, she could soon burst into a gallop from a standstill and back to a deadstop the second the aid was given. But she loves it, and so did Dusty during his rehabilitation process when we were looking for anything other than showjumping to stop him worrying at shows.

Pony fathers equipped with professional running shoes and grim expressions who drag often frightened children and nervous ponies through jam-packed lines of unled children are the biggest drawback as I see it. At most shows where led ponies are allowed to bulldoze their way to rosettes, it tends to be a bit of a shambles, but at the better organised Pony Club Mounted Games, you soon see why Prince Philip Cup ponies are at a premium and why their riders gain more confidence and learn more about control and balance than they had previously imagined was possible.

Judges and show-ring mothers

As with ponies, judges come in all shapes and sizes, from Lady Petunia Hornsby-Smythe of the Garth and South Berks, to Fred Gormless of the Dog and Duck.

A good judge is one who places your pony first. There will be times a-plenty when you will find yourself down the line and disagreeing with the judge's decision, but it is final nonetheless and you will win far more than a tawdry rosette at the end of the day if you hide your sorrows as your pony leaves the ring with the 'also rans'. Train yourself in front of a mirror at home to say, 'Oh bad luck darling, never mind, you were simply wonderful and there's always another day,' and at a show always wait until you get back to your trailer before throwing

Emma Thorp (age 14)

up. If the let-downs are tough on parents, just think of the damage done to children over the years.

We went to a local show recently where, as usual, all the competitors vanished as soon as a volunteer was required to jump first in the preliminary stage of the Nursery Stakes class. Unusually, the show was ripping through its schedule at a spanking pace, so as soon as we got to the collecting ring Sam was pounced upon by a frantic steward. With the speed of light, Dusty found himself before the judge and jumping the course before he had time to think about it, which is often the best policy for him.

It was one of his good days, and Sam was delighted with his clear round. We set about giving him a final polish for the showing section as, one by one, the curse broken, several other contenders came out of the woodwork to jump the course. After half an hour they were still short of the number of entries they had on their list.

Judge and steward wandered to the collecting ring to discuss with the organisers whether to close the class and proceed to judge the ones who'd been clear. Suddenly the judge whipped round and said to Sam, 'Have you jumped yet, dear?'

'Yes I have.'

'And did she go clear?' he asked the steward.

'I think so,' she said, 'What was your number again?'

The moral of the story is that being first to go, and the only pure white pony in a field of bays and chestnuts, and the only child not wearing a brace that day as you flash your teeth at the judge, and the only one in blue jodhpurs because your mother put them in the washing machine with a new pair of jeans, will not necessarily make an impression. In the final line-up we were second, but it didn't matter since the judge couldn't decide whether he was Arthur or Martha.

A couple of weeks later we were at a show where, more typically, everything was running behind time, and with both ponies doing different classes it was inevitable that they would clash. As Abbey rushed off to a showing class I rugged Dusty, now finished for the day, and left him with a haynet to go and see how Sam was progressing. I was amazed to see her coming towards me, with a snorting Abbey displaying a bunch of ribbons over her left ear. 'Mummy, we were fourth,' she said, absolutely delighted, 'and do you know what, we didn't have to do anything.'

It then transpired that as she rushed up to the ring, a straggly line was walking round judge and steward who were engrossed in conversation centre stage. Assuming that this was the start of the proceedings, our Sam slipped in, did the old heels down and shoulders back bit, and by the time she'd sorted herself out, three other ponies had been pulled in to stand in line. 'And we'll have the palomino next,' bellowed Mrs Elsie Dunkley, a rotund lady with a voice like a foghorn and more chins than the Hong Kong Telephone Directory, and of whom it has been said by members of the showing fraternity, that she is a better judge of a good gin and tonic than a good pony.

By the time I met up with Sam, the rest of the field had been sent out, cups and rosettes produced, and a lap of honour dispensed with, and then the penny dropped that she'd arrived at the *end* of the class and not at the beginning. This, and other highlights of the showing season where a pony placed first was owned by the judge and a child whose mother was the official timekeeper won two showjumping cups in succession on a pony whose reputation for competing against the clock is on a par with British Rail, convinced me that the governing bodies such as the British Show Pony Society and the British Show Jumping Association, who lay down rules for the sport, are worth the membership. When you do your local showing circuit for long enough, you get to the stage where you hardly need send for a schedule. The same old classes will be in the same old rings, with the same old judges wheeled out for yet another season. Will they still be showing there, with cobwebs collecting on their bowlers or floppy tulle hats, when I bring my grandchildren on a lead rein? I often wonder.

It's said that with age, more often than not, comes wisdom, and that being the case we were judged by one of the wisest of men at a popular Thames Valley show. He was judging the Working Hunters and had been doing so apparently since the Crusades, so there was a respectful silence around the ring as he solemnly lifted each pony's tail, bent down and peered up underneath with all the concentration of a fortune teller reading tea leaves. Gradually the impatient murmurings started.

'What's he doing down there?'

'It's his arthritis, he can't get up again.'

'He is eighty-six,' said the show-hunter judge who'd finished his classes days before in the adjoining ring.

'Well his predecessor was in her seventies and couldn't see the entrance to the showground without her bi-focals.'

'Is she still around?'

'Yes, they've got her stuffed in the secretary's tent.'

Just then, a parade with a brass band in the lead proceeded to encircle the show ground, and as it made its way around our ring, this dear old veteran stood to attention, then visibly left us for another world. He was smiling and as happy as a sandboy, conducting the band through its repertoire of tunes from days gone by.

'Lovely!' he pronounced, when they'd finished 'And shall we have some nice prizes?' to the sleeping children and ponies. 'That was a lot of fun, wasn't it?' he said to them as they went off, totally bewildered by a Working Hunter Pony Class where you were not required to jump. And so it was ... great fun. As I've always said, 'If you can't take a joke, you shouldn't have joined.'

LIKE NO BUSINESS I KNOW

The competitive element of showing tends to overshadow the social aspects. Yet even the ponies seem to enjoy meeting up with each other as horsy folk converge weekend after weekend, April to October. There are, after all, more than 2,000 shows in Britain each year from which to choose.

Out of the gossip and chit chat, come the real pearls of wisdom on subjects related to 'Pony Keeping Without Tears'. For every day spent with rainwater running in rivers down the inside of our jackets, a washing-up bowlful poured out of our boots, and wet socks on the dashboard as we drive home cold and shivering, there are many more where the sun shines and our only worry is whether we've brought the suntan oil to protect Abbey's nose – like all blondes, she tends to peel in tropical conditions.

It's at shows that you pick up invaluable home-tried and tested tips which you won't find in the reference books. At the start of last season, I feared that Dusty was bowing to old age at last – he was a trifle stiff behind as you might say. Barely noticeable, but it bothered me, and I mentioned it to a woman I'd only just met at the ringside. I knew about cod-liver oil for the relief of arthritic pains in both humans and animals, but I'd forgotten about the claims made for apple-cider vinegar until this woman recommended it, saying she'd had excellent results when she added it to the diet of her old pony.

It certainly seemed to do the trick, so now a flagon of cider vinegar sits on the shelf next to other natural health-giving aids such as Epsom salts which act as a cleanser of the bodyworks and cooler of the blood, especially beneficial therefore in early summer when the grass is very rich. It's good too for tired leg muscles at the end of the day, applied by soaking leg bandages in a strong Epsoms and water solution, its cooling action reduces swelling. Nearly all ringside conversations have taught me something, if only about the frailties of human nature. I've also learned never to comment on another pony or rider's performance unless it's to say something complimentary.

At one of our first shows, I was watching a Junior Jumping class, when a delinquent pony came in to the ring, napped at the sight of the first jump and bucked its rider off before charging back to the collecting ring. The child was in tears. The woman next to me was most unsympathetic.

'She should have used more leg,' she muttered.

'I don't think it would have made so much difference,' I said matily. 'That pony looks like a right little swine.'

'He's mine,' she said, totally dejected, and went off to pick up her screaming child.

Oddly enough, she returned, smiling and friendly, and we carried on talking, but I was pink with embarrassment and desperately thinking of something nice to say to make amends. Just then, an older child on a bigger pony came in to the ring, and this pony too napped and played up and finally reared and pranced about on its hindlegs. I was so relieved.

'There you are,' I said to my ringside companion, 'this will cheer you up, it's even worse than yours.'

'It's our other one,' she said, and with a look of despair went off as if to throw herself in the Thames.

There are showring mothers who are always a pleasure to meet. They roll up with their sunny smiles and their happy children on placid ponies, determined to make it a day to be remembered. There are others who wait grim faced and twitching round the ring, wallowing in the misfortunes of their rivals like the tricoteuse knitting round the guillotine as the executioner performs his grisly chore.

Sandra Crawford is an intensely ambitious pony mother who has led three children and an endless stream of tried and rejected ponies into battle, from the day they made their first gaff by having a pony in a drop noseband in a lead rein class, to the present day as she lives and breathes to produce a winner at Peterborough.

Today only one of her children and two of her ponies remain – the majority having failed to come up to scratch. The younger girl gave up showing the previous year when she left the showring in a flat out gallop straight through the boundary rope during a first ridden class, breaking an arm and a couple of ribs as the pony caught his legs in the rope and came crashing down on top of her.

The eldest child, a boy, was at odds with his mother's aspirations from the start. When he was four, his birthday present, which was a pure-white 11 hands Welsh Mountain pony, arrived in a protective wrapping of rugs, hoods, boots and bandages. When it was placed before him, the child was thrilled, thinking it was the tricycle he longed for, cunningly concealed. When it became obvious that the wrapping stayed on and his new toy was only unveiled for bathtime, before a show, or when parading in a ring, the novelty, even of owning the original pet of the invisible man, wore off.

A few years and two bigger ponies on, this boy finally cracked under his mother's constant lecturing coupled with the giggles and jeers of the girl competitors who outnumber the boys by about fifty to one. He begged to be excused to go to boarding school, where presumably he remains to this day. It's not something you talk about.

And so all hopes are pinned on Amanda, who is now into Working Hunter ponies. Her form has not been good lately and her mother is heading for her umpteenth nervous breakdown.

Now it always helps in a Working Hunter class to jump the course. If you don't get round in the first phase, despite the pony looking like a monument to Reckits Blue and Brylcreem, it's highly unlikely that the judge will be even remotely interested in your attendance at the second showing phase, which you've practised every spare moment.

Today the Crawford pony unceremoniously dumps its rider with a crafty stop at the second fence. The child has been conditioned like one of Pavlov's dogs. When she gets a rosette, an ice cream, chocolate or other treat will surely follow, so having crashed to the bottom of the popularity stakes, she's distraught to say the least. Failure equals rejection by her mother, and a frosty homecoming from the rest of the family. Not so much because they care that they're related to a loser, but because they too will suffer mother's black mood until another victory appears on the horizon.

Mother is inconsolable throughout the public inquest into today's disaster, and wailing like a banshee. 'In the seven years that pony's been doing these classes summer and winter without a break, it's never stopped. It MUST be the rider. Why, when I've spent my last penny on ungrateful children, can they humiliate me by letting rubbish get places in front of us, destroying our reputation and giving everyone a good laugh no doubt.' The ringside crowd who've heard it all before, drift away, bored with the display of vanity and bad sportsmanship, the foul temper and whatever else it is that perhaps a psychiatrist could explain.

This is a rare form of child abuse, the need to bask in the reflected glory of the child, who, through no fault of her own, wasn't born with what it takes to make her mother a success. One day when that child grows up, possibly warped and emotionally stunted by her turbulent days in the showring, she'll reflect on the childhood her mother endured as one of the 'also rans'. When the mother grew up and the man of her dreams failed to live up to her expectations, there was still a chance to bring, through the children, some of the colour and excitement she craved in her pretty average existence.

And perhaps kindly, because she always wanted badly to please her mother and to love her, the child will tell herself, 'She only wanted the best for me.'

YOUR COMMENTATOR AT THE RINGSIDE

If equestrian pursuits were once the privilege of the rich and famous, today nothing could be further from the truth. At any pony event, the aristocracy not only graciously mingle unabashed with the hoi polloi but get together to share secrets, complain about the organisers, or just have a quiet bitch about somebody. Yes, 'If wishes were horses beggars would ride' was the old saying – and nowadays their wishes have been granted. Many's the pony who owes his subsistence to his owner's hand-out from the Department of Health and Social Security.

At one show a well-known multi-millionaire parked his new Rolls at the ringside and watched delightedly as his daughter won her class. For all the showground to hear, he then yelled and guffawed on the car telephone to Mummy in the Bahamas, instructing her to open the champagne.

Meanwhile, a few yards away, a scruffy-looking woman was emptying her brood from a ramshackle vehicle, which appeared to be held together with rust and pieces of string. But the children and ponies were immaculate, they could have graced the pages of *Horse and Hound* or any Society publication where they could have been seen but not

Toby George (age 14) and Dulcie George (age 10)

heard. At the previous show they attended, the eldest boy had been heard to mutter an obscenity upon being placed down the line, to the horror of the attendant officials. Their mother, conscious of the fact that it would be unwise to encourage such loutish behaviour, was giving them a last-minute pep telk on etiquette.

As they departed for their various classes, she lowered her voice to a shout and reminded them, 'Don't any of you little buggers let a judge catch you swearing!'

MADAME DEFARGE

Equestrian sports aren't any different from others where the object is to win. It's no sin to want other people to admire an animal you think is the tops, but it takes a special kind of person to treat success with the same disdain that they treat failure. Ponies, generally speaking, belong to young and vulnerable competitors who must be forgiven the occasional displays of pride or jealousy. The behaviour is far less excusable in adults. People who have what it takes to get to the top can afford to be generous and helpful to those who are less fortunate. Those who aren't in the same league, or who are afraid they never will be, destroy the confidence of those around them by using any means. So it is with the business of showing ponies. Adults who go around crowing about their successes and having tantrums when they don't win, spoil the sport for others and would be well advised to give it up.

Each week at local shows as the wagons roll on to the showground, you'll hear mutterings as the 'pot hunters' arrive. The ponies and riders who are way out of their class at local events, having already qualified for Peterborough and the Royal International, now come to try their hand at the plastic trophy and bag of pony nuts on offer today for Best Pony at Goosey Common. And the small child who has hacked to the show on the specimen she's rented for the day from the local riding school, is about to have her dreams of a rosette (of any colour so long as she gets to keep it), shattered by the arrival of the professionals. But if you can't beat them, at least learn how they win.

One day when Abbey started to behave as though harbouring tarantulas in her ears, shaking and nodding, her mind miles away from the business in the ring, it was one of the professionals who volunteered to inspect her tack after she'd been put down a couple of places following her individual show. She suggested that her bit was lying too low in her mouth, and corrected it, whereupon the pony was much happier. When in a subsequent class our pony was placed above hers, to our surprise, because her pony is a cracker, she was the first to say well done.

It will not go down well with an experienced competitor, who takes the matter of winning seriously, if you go up and ask what exactly she feeds her pony on, just as she's about to enter the ring, any more than it would do to ask Nana Mouskouri for the name and address of her optician just as the orchestra signals her entrance on stage. However, by choosing the right moment – as they sling the cups in the lorry or finish counting their winnings, or drop the bottle of whisky into the judge's car – most pony folk will accept your desire to imitate as the sincerest form of flattery and give you any advice they can.

BSPS

Showing classes at many of the bigger events are affiliated to the British Show Pony Society. Affiliated ponies are those owned and ridden by members of the BSPS. Membership runs from the first of January each year, and a book of rules (sent to you together with an application form) should be read thoroughly before you proceed with the associated expenditure, such as obtaining the necessary height certificate under the British Equestrian Centre's Joint Measurement Scheme. This ensures that a pony is exhibited strictly within the height limits for the class in which he is entered. Only veterinary surgeons on the official JMS list can measure a pony for you. Furthermore, under a new ruling which came into effect on 1 January 1987, all requests for measurement must be sent to the JMS office, by the owner/agent. You have to complete a form giving details of the animal and his ownership and giving the names of two official measurers in your area (which you select from a booklet sent to you along with the form).

The JMS will forward this information to the appointed official measurer who will telephone you to arrange an appointment. In the old days, these OMs came to *you*, and measured the pony on a suitable piece of level ground, but now measurers have to use official registered JMS pads to which you take the pony, where he is then accurately measured and a certificate is issued which will be valid for twelve months from the date of measurement. It is required that the pony is unshod on the date agreed for the measurement. Care will be taken to ensure that he's completely relaxed before the measuring takes place – ponies can easily 'grow' by a couple of inches simply by being nervous of strange happenings around them.

Because it is desirable to have ponies at the top of the height limit for their class, sordid tales abound of ponies being made to stand for ages with heavy weights across their shoulders, or of having pins stuck in their withers to make them tender so that as the measuring stick is lowered, the pony shrinks from it in anticipation. It would be nice to think that such stories are entirely fictional, but since outwardly normal people will leave ponies with their heads tied back in draw reins for hours on end to improve their headcarriage for a showring, wrap them up in polythene to make them sweat profusely to lose their winter coats, then toss them out into cold fields and forget them when the showing season is over without a 'roughing off' period to grow the coat again, who knows what goes on in the name of sport. It was E. Hartley Edwards who made the observation that there are no problem horses, only problem owners. I've met more than one vet who agrees with that.

Back to the Height Certificate. Ponies are given an annual certificate until the age of six when they are eligible for a Life Certificate. A pony six years and over at the time of joining, will still only get a certificate for one year and then must be measured again for a Life Certificate. The BSPS naturally deplores any malpractice in the training or exhibiting of show ponies. The organisation and the judging at affiliated shows are generally of a higher standard than at gymkhana level, as a strict set of rules must be adhered to and members of the panel of BSPS judges are carefully chosen because of their vast experience with ponies.

A pony does not have to be registered with a breed society to be eligible for membership. Once joined, producer, pony and rider will be aiming to qualify for the Horse of the Year Show at Wembley in October for riding ponies between 12.2 and 14.2 hands, and for the BSPS Championship Show at Peterborough, if their interest lies in Leading Rein, First Ridden or Working Hunter Pony classes.

Further information and details of fees from:
The British Show Pony Society, 124 Green End Road, Sawtry, Huntingdon, Cambridgeshire (Tel. Ramsey (0487) 831376).

Other Useful Addresses:
The National Pony Society, Colonel A.R. Whent (Secretary), Brook

House, 25 High Street, Alton, Hampshire GU34 1AW.
The British Horse Society, British Equestrian Centre, Stoneleigh, Kenilworth, Warwickshire CV8 2LR.

Showjumping

At the level where young children compete, there is a marked division between those who show and those who showjump. We started off with the intention of showjumping Dusty, only to discover that it frightened him out of his wits, and we have to assume that he has had an unhappy experience in jumping competitions since he is in his element on cross-country courses and enjoys showing off in Working Hunter classes.

If you have ever watched, as I have, at the smaller local shows, grown men and women beat a pony before it enters a ring, to 'get it on its toes' ready to dash forward and away from the whip as the bell goes at a jump off; if you have seen child riders with spurs and whips scream and batter at their ponies to get them to go faster and higher, with their parents at the ringside vocally encouraging them, you will not be surprised as to why so many old jumping ponies, their spirits finally broken, end up a sorry sight. It must also account in some way for the growing popularity of Working Hunter Pony classes, where good manners, good riding and an ability to jump are required.

Thankfully, with affiliated jumping, the British Show Jumping Association requires that riders and ponies must be registered by the Association and abide by the rules designed to encourage riding skills and to prevent misuse of the ponies. Membership details from The British Show Jumping Association, British Equestrian Centre, Stoneleigh, Kenilworth, Warwickshire CV8 2LR.

WHAT PRICE A TROPHY?

What, at the end of the showing season, is it all going to get you? You could probably have bought yourself a nice gold watch for the amount of money you've paid for the rosettes, if you tot up everything from the entry fees to the small lake of petrol needed to take you round the various events. And the cups, which you may have (collecting dust and tarnish) for less time than you thought, are going to cost pounds for return postage and engraving the name of this year's winner. Look at the small print before you sign for them, you'll notice that the worry of replacing them should they go missing won't last for long. By the time the long-life silver polish has worn off, the secretary will need them back. If the show is in August, this could mean a return by May – deduct the time it spends at the engravers and you'll hardly see it.

I rather like the idea of giving children even a small trophy they can keep when the beautiful cup has gone back, and their pony days may be over. Some organisers who've adopted this idea, obviously see it as a financially viable proposition....

At one show where Abbey surpassed herself, Sam was presented with a plywood block with a 'gold' statue of a pony on top, which was hers to keep for winning, and the little chap who was runner up pocketed a brown envelope containing £4 for second place. It was only when we popped into our local sports shop, where they do a roaring trade in trophies for engraving, that we noticed our 'gold' statue for £2.50.

There are competitions where an outright winner is chosen at the end of the season on a points system. This means contenders have to attend every one of the preliminary shows and enter as many classes as they can possibly stretch the rules to get away with (Lead Rein pony under 15 hands etc.) in the scrabble to be champion on points. By the time the cup is presented they could not only have bought it outright but had a fortnight for two in Majorca.

I often wonder if the original owner of Priorswood Dusty Boy was a Rockefeller or Vanderbilt. It costs an arm and a leg to engrave a long name, so I would recommend to anyone with a youngster displaying showing potential, to give it a name like Jo or Di. Apart from the money you'll save, show secretaries, who have to fill in entries on the day and commentators who can never get names right, will love you.

The husband of an acquaintance christened his Irish Hunter 'Business', so that his wife or secretary could truthfully say he was out on Business when anybody called. His one and only day as a showjumping competitor was marred not only by the fact that he fell off and broke his nose crashing through the double, but by the commentator completely putting him off his stride by pronouncing the horse's name Boo-sy-nus. Understandable, I suppose, if you look down the list of runners at some shows. Anything that hasn't got a name like Fluorescent Bubbling Sparkle must be working down the pits the rest of the week.

NAMES

The naming of children is almost as interesting. Although the majority of native children at pony events seem to be divided into Emmas, Sarahs, Samanthas and Joannas, mothers who aspire to the stature of thoroughbreds take care in the christening of children to name them after the Lucindas, Virginias and Annelis of the upper echelons of equestrian events. I remember a conversation which my down-to-earth Scottish father was having with his granddaughter on her return from Pony Club Camp a few summers ago.

'And were there nice girls in your dormitory?' he asked.

'Lovely thanks. I shared with Lucinda, Imogen, Belinda and Florence, and we managed to keep masses of tuck for midnight feasts.'

My father turned and looked wearily at me. 'Good God, lassie,' he sighed, 'Nae wonder we've got striking miners!'

9 TIPS FROM THE TOP

Jenefer Hall (age 12)

DAVID BROOME

International showjumper, former Pony Club member and erstwhile child competitor in the showing ring. Brother of Liz who married Ted Edgar.

'In truth we only did showing for the money, if we could get ten bob for third place, it was worth a go, but that was the only reason. We were a bit rough and ready and it was all too prim and proper for my liking. Plus your success rests on the whims of a few judges – showing is a small world and, if you notice, the same ponies win time and time again. With show jumping, if you clock up four faults, it's there for all to see and no arguments, the system of judging showjumping is a much fairer one to my mind.

'I don't approve at all of small children on speedy little ponies driven like crazy over fences for the amusement of adults – which is all it is. It isn't doing anything for the rider or the pony. When we were young they had a different, and I think a better, system. We had to go three rounds to get the winners, rather than a second and final one against the clock, some of the little speed merchants in any case can't cope with height, and it encouraged a better standard of riding.

'We never had any money, but as kids we had the advantage of a farm with space and facilities to keep ponies. The ones we had certainly didn't cost much to buy, we had to make them good, which was much more fun. Children nowadays have more help than we had to get to the top. It's now such a popular sport, but my advice would be take it easy and don't lose sight of the fact that it's meant to be enjoyable. There's a lot of pressure put on kids to aim for the top, which they can't cope with. Jimmy Connors has the right idea in tennis. He says you can only concentrate on gaining one point at a time, and that's true with riding. Aim to do your best at your particular level, whatever that is, and the rest will naturally follow.'

STELLA HARRIES

Trains top-class horses, ponies and their riders at her Little Barn Stables near Bracknell in Berkshire.

'You can be lucky and buy a good animal with potential for a fairly modest sum and bring it on yourself, providing you have the necessary experience. The trouble lies with people who don't like to admit they don't know enough about it. My advice to an inexperienced parent when buying a suitable pony for a child, would be to put your trust in a knowledgeable expert. Reputable professional dealers belong to quite a small circle. If they don't have something to suit you in their yard, they will contact someone in another part of the country who has, they won't just offer you the first thing that comes along. You may pay a little more initially for this service, but you will at least have some sort of a guarantee that the pony will suit your needs. Unfortunately you don't always get what you pay for and the worst thing to my mind are the exchanges of ponies between private individuals. Often a mother who has a bad animal which has been no good for her own child will lie through her teeth about its true qualities in order to pass it off to someone else.

'Most good animals are made by good partnerships, they will go places and do great things for a sympathetic child who treats them as a well-trusted friend. I don't think it should be too much of a risk to buy a pony for a novice which has previously belonged to a good rider. After all, if the pony has been taught well, it can teach the new rider good habits and give confidence, because it's willing to do everything that's asked of it.

'I don't think regular lessons are vital, but it does depend on the individual. A bright child will remember everything she's been taught at an occasional lesson and apply it when she needs to. Getting a proven top-grade instructor is more important than having many sessions with someone who has an inflated opinion of their teaching abilities. The best advice I was given as a child rider, which I still pass on to my pupils today, is this, "You never stop learning. The day you think you know it all, give up."

MALCOLM PYRAH
International Showjumper who is now enjoying his five-year-old daughter's interest in ponies and riding.

'Basically she likes to do what her parents are doing. We ride, so naturally she does. I suppose you could say we encourage her, but we are always careful not to push it. She doesn't have any formal teaching at her age, she's at the stage where she's just getting confidence and it's all good fun.

'When the time comes and she wants to learn properly, I'll see that she does. I do think lessons are very important, far more important than we tend to regard them in this country. When you travel, you see how much more attention to groundwork foreign competitors consider is necessary. Here, we are preoccupied with instant success, children get on and start galloping around without a sound basic training, and of course their riding, and consequently the pony, suffers. I think we should put more emphasis on style of the rider and style of the jumping, not just who knocked a fence down and who didn't.

'I enjoyed my early days as a Pony Club member, I think it's a good organisation for kids to get a chance to compete at a decent level and it also gives the non-showy ones a chance. The trouble with show riders is that they can perform beautiful little displays in a ring, but wouldn't know how to put a bridle on. I wouldn't say that money can buy showing success, but it certainly helps. On the other hand, we started with ordinary gymkhana ponies, got a taste for winning and progressed from there. I do think you have to want to win to get anywhere, and by that I mean a child has got to have a competitive instinct to be noticed. Certainly there's enough competition amongst the parents. In fact one of the biggest drawbacks to a child's progress and pleasure is the constant pressure from grown-ups. The egos of the parents shouldn't ride on the backs of children.'

JENNY PITMAN
Britain's most famous and most successful lady racehorse trainer. Mother of Mark, who always wanted to be a jockey, and Paul, who preferred golf.

'I never pushed my boys to do anything. Your children are only on loan to you for precious few years until they go off and do their own thing, and above all I want mine to be happy. Having said that, from a selfish point of view, I wanted them to be interested in sports, because I am too. If I'd had girls, I would have been only too pleased to help them lead the life I've had – I now get paid for looking after horses which is what I enjoy best.

'My father sat me on my first pony, bought at a sale, when I was hardly past the toddler stage – he knew a lot about ponies and a lot about life. I even went to school on my pony, I was the one in the front row wearing jodhpurs in the school photograph, and we played Cowboys and Indians until after dark, listening to the ground for hoofbeats to tell us the others were coming.

'There was no question of shirking the work involved for their welfare, but we didn't want to, it's all to do with the standards you're brought up with. Remember a pony is flesh and blood, not a bike you can put away when the weather's bad. You must ask yourself if you can afford, not just to buy a pony, but to give him the best of everything. The best quality food, regular visits from the blacksmith, the best of health care, and take time to understand how he feels.

'I think girls and women understand the needs of a pony or horse more, but there's still so much ignorance in areas such as feeding. People lash out on fancy tack for a show and economise on the feeding to pay for it. You can't lay down hard and fast rules for feeding, but you must understand the principles. A cold pony, even one who appears to be getting a lot of food, is just going to lose the benefits in burning up that fuel in an effort to keep himself warm. If you make sure he's warm and also feed wisely, you can keep weight and condition on the pony and subsequently he has more resistance to illness and more stamina to do his job.'

STEPHEN SMITH
Son of famous father, Harvey, and a successful showjumper in his own right.

'I wouldn't say my brother Robert and I were pushed to ride, but my father would have been disappointed if we hadn't, I'm sure. I think, left to myself, I would have become a footballer, I never really bothered much about ordinary riding. I hate hunting and I can't understand how you can just go round boring roads and fields, even today I don't like schooling in a paddock, just the riding at shows. That was the only thing that got me interested when I was a kid. Once I could win a class or two, then it was something else. You get hooked on winning like a gambler does.

'I don't think parents can make good teachers, at least ours didn't. We just got stuck on and shouted at, but I must say we got the chances which a lot of kids would have done anything for. I've never had any regard for the showing side of competition work. I think it's a complete and utter waste of time from a rider's point of view. There are a lot of highly strung ponies which some of the kids don't even get to sit on until a show, then they have to perform in front of a whole lot of bent judges, and can't understand what they did wrong, why they didn't get picked. In show jumping at least you can understand what it's all about.

'You have to be kind and firm with a pony to get the best out of him. If he's punished when he's wrong, then equally he should be praised when he's right. Trouble is half the people who own them don't know

131

the difference, and they think that having a soft attitude shows they're animal lovers. It gets dangerous when it's sentimental. You see kids getting smashed all round the place by ponies that are real little gangsters, the kid loses his nerve, the pony ends up shot and everybody says, 'Oh the poor pony,' when all it needed was a good thrashing when it was starting to get the upper hand. Good ponies make good riders, and the best one you'll ever have is a novice that hasn't been ruined by somebody else's bad handling. You've nobody but yourself to blame if a novice turns out no good. My advice to anyone starting is, "Don't, it's nothing but headache and aggravation." I often wish I'd been a footballer.'

GEORGE EDWINS

One of this country's leading trainers of showjumping juniors at the Foxlynch Training Centre near Marlborough, Wiltshire. His daughter, Amanda, didn't ride at all until she was ten, and father only began her serious training when she was thirteen. Within a year she was Captain of the British team which won the Nations Cup in Sweden in 1977. Of Amanda Edwins' success, George says, 'The thing I always admired in her was her relationship with her ponies out of the ring. She wouldn't have anyone else attend to them. Consequently she could tell immediately when they were out of sorts, foresee any problems, and attend to them before they got worse. Ponies can get out of bed on the wrong side too, and if on the day of a show she spotted they were a bit stroppy, she would give them a bit of work in the paddock to settle them before we set off. She's really tuned in to them whether on their backs or doing them in the yard.

'It's for this reason that I think grooms are the biggest curse to children and ponies. The children are handed the pony to get on and they can never form a real understanding of what makes him tick. Another pain is mothers who fuss and fret and regard the pony as their very own, with the child borrowed to ride it. I've seen mothers be physically sick with nerves at shows, so how can a child or pony be unaffected? Without a doubt, a lesson I give to a young rider is far more productive when the parent goes off and leaves us.

'To get a successful combination of pony and rider you need someone good on the ground to assess them and let them develop their own style – that's the secret of good teaching. In America the trainer is highly regarded. Here, we haven't got many that are good enough to produce talented kids and ponies, but if you're lucky enough to find one, always pay for the best tuition you can afford and never grudge a pony the good grub he needs to do the job. In July and August when they are doing so much travelling, which is more tiring for them than the jumping, we add a bottle of Guinness and three eggs to their food as a tonic.'

DOUGLAS REID
British Show Pony Society Panel Judge and Editor of the North West Regional Equestrian magazine *Hoofprint*. Anyone who regards judges as ogres and crooks should meet this lovely man who tries to put his smallest competitors at ease by an informal preliminary chat – even walking the course with them, if possible, and advising them on the best way to tackle it. Once they regard him as a kindly grandfather figure, the event becomes less of an ordeal and more of an occasion to remember. There are many judges and judgesses who would do well to study his methods.

'The most important advice I would give to any competitor is, in a word, RELAX! There are two related reasons why a rider should be completely relaxed. Firstly, the expression "stiff with fright" means, quite literally that tension in the mind leads to tension in the body and a stiffening of the muscles, which in turn leads to clumsy, uncoordinated riding. Secondly, ponies and horses are telepathic, and any uneasiness in the rider's mind will be immediately transmitted to the pony's mind. What is worse is that whilst he senses the rider's unease, he has no means of knowing what is causing it, and will consequently be "spooky" about everything.

'The rider who enters the ring with "the ring of confidence" will impart that confidence to the pony, and both will give a far better performance than the tense, stiff rider with a pony too full of uncertainty to concentrate on his job.'

Author's footnote: The best evidence I ever had of the Douglas Reid philosophy, was in the Middle East where I and a group of British women who had had assorted riding experience, were permitted to take the Arab racehorses on their daily exercise sessions down to the beach and across the desert. We had a merry time being carted off or dumped in the sand as, not infrequently, stallions attempted to co-habit with the mares regardless of whether there were passengers aboard ... a gelding is something unknown in these parts. These gorgeous-looking beasts gave us quite a time of it, bucking, rearing and galloping up the familiar stretches where their grooms encouraged them to go flat out. It was impossible by this time to stop them, you simply held tight and prayed fast. Eventually, worn out with the effort of staying on and keeping a stiff upper lip, we arrived back at the stables with spirals of steam rising from our bodies and the horses, with fixed smiles on our faces as we related to the grooms the lovely time we'd had, and toddled off with legs like jelly to sink in a bath of Pimms No 1.

Then came the day of the annual parade through the town with its military bands, tanks, low-flying air displays, gun salutes and other goodies for highly strung horses. 'Our' horses were due to take part so we went along and positioned ourselves in the grandstand to see the fun. We couldn't pick out our fiery steeds at all, until an excited groom spotted us and pointed in the direction of a bunch of beautifully turned out, but apparently stupefied animals, plodding along on long droopy reins. They were ridden by tiny barefoot little boys, no more than about ten years old, who were having a smashing time grinning and waving at their friends in the audience. One little chap found it easier to avoid the restriction of his long white cotton robe which was his only riding habit, by sitting cross-legged as he would do at home, so he did.

We drove to the stables to watch them return the three miles home, a gang of merry urchins on perfect family ponies. The children scrambled off their backs and ran to queue before the head groom for payment. The brave British *memsahibs* just stood there, stunned into silence and feeling as sick as parrots.

'How long have you been riding?' asked one of our lot at last, when a fellow no higher than her knee came over to say hello.

'This is my second time,' he answered with a beaming smile, 'but my brother's been taking them to the last six races and he told me the money was good.'

'Don't you ever feel frightened?'

'Of what?'

'Oh nothing ... What are you saving up for?'

'My own motorbike,' he said, leaping on to the back of his brother's bike and roaring off in a cloud of dust across the desert.

MR GORDON GORSE

If this sensible gentleman is now the late Mr Gordon Gorse, I'm sorry to hear that, because his book, *The Young Rider* (published in 1926, before everything about keeping ponies became so complicated) was given to Sam as a reward for selling raffle tickets at a recent charity auction, and it is a gem amongst pony books. As I have no way of contacting Mr Gorse, except perhaps through a medium, and I have tried to trace members of his family in vain, I have taken the liberty of including a few of his tips and observations based on stable talk in the earlier part of this century.

On the pony: A well-broken pony has no faults, but he can easily pick up bad habits. He will always do well if he understands.

On good riding:
'Your head and your heart keep up
Your hands and your heels keep down
Your knees press into your pony's sides
And your elbows into your own.'

On the groom: Remember, in *Black Beauty*, John the groom had special balls, which he fed to the horses and ponies? These 'balls' were made up of patience and gentleness, firmness and petting, one pound of each mixed with half a pint of common sense and given to the pony every day.

On etiquette: Well-trained ponies will obey a word spoken quietly, and the noise of shouting which proceeds from some second class stables marks them for what they are.

On stabling versus living out: It is expensive to keep a stabled pony, somewhere between 25 and 35 shillings per week. He will be in hard condition and up to any amount of work. In this way he will be superior to the grass-fed pony, but in no other.

On choosing a pony: A pony who is too keen scares his rider. It is the worst possible fault. A slug is dull to ride and teaches a child bad ways. He learns to pull the pony around as if it were a cow!

On colour: Chestnuts are inclined to be excitable. A grey pony is usually a good one and often very beautiful, and his dark eyes add to the beauty of his head. Greys go white as they age. There are, I believe, no ponies white in their early years. Queen Victoria's cream-coloured horses were very handsome in procession, but horse lovers have never cared very much about them. They had pink noses which reminded one of ferrets.

On dealers: Personally I prefer buying from a dealer. A private owner has often exaggerated ideas of the value of his possessions.

On fly repellents: Jeyes Fluid sponged over ears and neck is an excellent fly driver.

On road safety:
'The Rule of the Road is a paradox quite
Both in riding and driving along
If you keep to the left you are sure to be right
If you keep to the right, you are wrong.'

10 BANKRUPTCY AND BEYOND

Deborah Hall (age 14)

The day will finally arrive when the pony mother graduates with all the hallmarks of the genuine article. Nails bitten to the quick with worry, she is prematurely grey and permanently overdrawn. Her husband is taking more than a passing interest in the glamorous redhead who lives in a home which isn't a cross between the Pony Club Headquarters and a second-hand tack shop. At least a pony doesn't complain that she smells like a jute rug, has hair like Topsy and hands that would be rough on a builder's labourer.

That sweet little pony – which cost less than the week in Paris they once meant to have – has now run up ten times as much in just his first year with the vetting, insurance, vaccinations, shoes, tack and rugs, food, trailer, petrol, Pony Club subs, and entry fees, riding clothes for the jockey and lessons to help them synchronise their performance. This is assuming you have your own place to keep him. If not, add on anything from £5 to £35 for his lodgings.

'Do you know what it costs to keep a pony?' demanded a friend not so long ago, when I asked why she didn't get one for her pony-mad daughter. This child, like many others, is being exploited by her local riding school putting in too much work in exchange for too few rides, and all because she's crazy about the animals. Her mother has no interest whatsoever in ponies, which is the major obstacle. The next problem is that mother, being a maths teacher, finds it necessary to do sums all the time. I was dumbfounded at her question, and of course I don't know the answer. It's as daft as asking how much does it take to fill a bottomless pit? Certainly it isn't cheap, and I don't know anybody who keeps ponies for their own pleasure and makes a profit from them, however good they are. But is it worth it at the end of the day? – that's what you must ask yourself. And frankly many are the days when I have answered, 'No, why am I doing this?' Days like the one spent getting plastered at the hospital with a smashed-up wrist, or when Abbey has stepped back on to my foot at a show and I have to drive home twenty miles with an ankle so swollen I can't get a shoe back on. Any person who tells you that a pony will never stand on you if at all possible, simply hasn't lived with Abbey. She can stand on anybody within a ten-yard radius.

Then there are the days lying on the osteopath's expensive couch with a displaced disc through lifting up the heavy ramp of the horse box. Days of getting hopelessly lost trying to find shows or Pony Club events which are signposted by a dirty postcard tucked into a grass verge on the opposite side of the road. You would need a Sherpa to find some of the venues for local shows.

And then there are the really dreadful days. Not just the day we found Dusty apparently choking to death because (our vet's guess) he suddenly looked up at something which startled him, and a bit of grass went down the wrong way. And not the fact that he did a similar start resulting in a horrendous looking eye injury where he jabbed himself on a twig. But the worst day of all, after we'd had an overnight storm and a tree had fallen over a neighbour's fence and into our paddock... Standing next to it, was our Papi, lame.

To this day, we can't be sure why, but he always was a great acrobat and a climber, he loved to stand up on his hind lgs, resting on the stable door or on poor Albert our blacksmith, or whatever else took his fancy. He may have climbed on to the tree....

My biggest mistake was not calling a vet immediately. Well-meaning friends who knew much more than I about horsemanship assured me it was just a tiny nick in the heel and that he'd be as right as rain by the next day – it certainly looked that way at the time – but I should still have relied on my instincts and summoned immediate veterinary help. There is so much to learn about the care and handling of ponies and I really knew too little about them to take on the responsibility of ownership in the first place. I was to pay dearly for that, because it wasn't the cut that was the problem.

Next morning he was still lame, and when a vet arrived he delivered the dreadful verdict. I had expected him to say that Papi couldn't be ridden for weeks to come, but when he said that he had a fifty-fifty chance of full recovery with a stretched tendon, I just couldn't take it in. I felt so guilty, so full of shame that he'd depended on me and I'd let him down.

For five weeks he was stabled with only hay and small feeds of bran mashes to eat, and had poultices applied twice daily to his leg. Chris and Lorraine were marvellous, lending a practical hand when it came to nursing him around the clock, and moral support through what seemed an endless wait. The children remained optimistic, as children do when they are very young and grown ups are supposed to know best, and vets are magicians and sick ponies don't look like Papi, with eyes like buttons, a glossy coat, and so pleased to see everyone.

When the vet announced there wasn't any point in him coming again Papi would either get better or he wouldn't, in desperation I tracked down a faith healer. This man worked as a scientist by day at a research centre in Oxfordshire, and offered to treat Papi daily in his lunch-hour if I could collect him from work and take him back. I couldn't believe his kindness. Papi loved him, and his visits gave me new hope. The pony lay down, closed his eyes and slept as hands were laid on the injured leg, it was very touching to see. Whether people believe in spiritualism or not is entirely their own affair, but this treatment certainly did no harm and anything was worth a try. We'd been told that by the sixth week we'd know, conclusively, the extent of the injury.

I still couldn't accept it when told that he would have to be destroyed, and of course, wanted another opinion. I was, at the time, working on BBC Radio 4's *Profile* with Jenny Pitman who recommended her marvellous vet, Barry Park, making me promise that if there was nothing more to be done, I would let the pony go.

Looking back, it seemed I'd always known what the outcome would be, and there was simply nothing left to do but say goodbye to him. I was haunted for ages by the look on his face as we loaded him in a trailer to take him to a nice quiet field. Watching me, bright and trusting in happy anticipation of what was to come at the journey's end.

I thought my heart would break when it was all over, he had so many useful years ahead of him. Wilma, who'd raised him from babyhood, tried to make me feel better by talking of the lovely life he'd had, unlike so many other little ponies. He was never short of food or love, he didn't know what a stick was, and he lived for being out with little people and doing his job, but he would have hated life as a permanently lame pony stuck in a field.

So often I remember Papi with great sorrow. On such days, I find the best thing is to go out and talk to 'the boys'. I look at Dusty and I can't believe our luck in finding him. At Abbey, the ugly duckling who turnd into a swan, or at least a working swan, and whose sweet existence provided the cure for Sam's unhappiness over Papi. And then there's dear old Gay, who's managed not to drop me since we met, and who asks for so little in return for her good and patient service ... and then I'm sure it's worth it – every bit.

RECOMMENDED READING

Akrill, Caroline, *Showing The Ridden Pony* (J. A. Allen, 1981)

British Horse Society, *Manual of Horsemanship* (BHS Publications)

Codrington, Lt.-Col. W. S., *Know Your Horse* (J. A. Allen, 1974)

Edwards, E. Hartley, *Saddlery* (J. A. Allen, 1985)

Giles, Brian, *So You Think You Know About Horses? Questions to test your knowledge* (Stanley Paul, 1985)

Oliver, Robert, *Showing Horses and Ponies* (Pelham Books, 1985)

Parsons, Derrick, *Do Your Own Horse* (J. A. Allen, 1977)

Rees, Lucy, *The Horse's Mind* (Stanley Paul, 1984)

Roberts, Pamela, *Teaching The Child Rider* (J. A. Allen, 1973)

Rose, Mary, *The Horsemaster's Notebook* (Harrap, 1977)

INDEX